Interaction of Radiation with Matter

Interaction of Radiation with Matter

Hooshang Nikjoo

Shuzo Uehara

Dimitris Emfietzoglou

CRC Press
Taylor & Francis Group
Boca Raton   London   New York

CRC Press is an imprint of the
Taylor & Francis Group, an **informa** business

CRC Press
Taylor & Francis Group
6000 Broken Sound Parkway NW, Suite 300
Boca Raton, FL 33487-2742

© 2012 by Taylor & Francis Group, LLC
CRC Press is an imprint of Taylor & Francis Group, an Informa business

No claim to original U.S. Government works

Printed in the United States of America on acid-free paper
Version Date: 20120412

International Standard Book Number: 978-1-4398-5357-3 (Hardback)

Library of Congress Cataloging-in-Publication Data

Nikjoo, Hooshang.
   Interaction of radiation with matter / Hooshang Nikjoo, Shuzo Uehara, Dimitris Emfietzoglou.
      p. cm.
   Includes bibliographical references and index.
   ISBN 978-1-4398-5357-3 (hardback)
   1. Particle tracks (Nuclear physics) 2. Ionizing radiation. 3. Radiobiology. 4. Materials--Effect of radiation on. I. Uehara, Shuzo. II. Emfietzoglou, Dimitris. III. Title.

QC793.3.T67N55 2012
539.2--dc23   2012008917

**Visit the Taylor & Francis Web site at
http://www.taylorandfrancis.com**

**and the CRC Press Web site at
http://www.crcpress.com**

Contents

Section II

Section III

Section IV

Preface

The subject of this book is primarily the physics of the interactions of ionizing radiation in water and the Monte Carlo simulation of radiation tracks. The book explores the subject from an elementary level and progresses to the state of the art in our physical understanding of radiation track structure in living matter. Section I of the book deals with the elementary knowledge of the radiation field. In Section II we explore the cross sections for electrons and heavy ions—the most important information in need for the simulation of radiation track at the molecular level. And in Section III we discuss in some detail the inelastic scattering and energy loss of charged particles in condensed media with emphasis on liquid water. Section IV provides a large number of questions and problems to explore the subject of this book.

The book was partly designed to be used as a textbook in radiation interaction courses. More generally, we hope the book becomes a platform for education in this topic at the master and PhD levels for medical physics, health physics, and nuclear engineering students.

We express our gratitude to all those who have given us help with the preparation of the book. In particular, we like to thank colleagues at the Radiation Biophysics Group of the Karolinska Institute, Peter Girard, Thiansin Liamsuwan, Reza Taleei, Tommy Sundstrom, Lennart Lindborg, and Krishnaswami Sankaranarayanan for their help.

About the Authors

Hooshang Nikjoo is professor of radiation biophysics at the Department of Oncology-Pathology, Karolinska Institutet. He obtained his BSc (hon) in physics and PhD in radiation physics at the Polytechnique of South Bank London. Previously, he worked for many years at the MRC Radiobiology Unit, later renamed the Genome Stability Unit, Harwell, Oxfordshire, UK. His scientific research interests encompass computational approaches in molecular radiation biology, including Monte Carlo track structure methods, modeling DNA  damage and repair, and estimation of radiation risk in humans from exposures to low levels of ionizing radiations using a genome-based framework.

Shuzo Uehara is an emeritus professor of physics at the School of Health Sciences, Kyushu University. He graduated from the Department of Physics of Kyushu University and received his MSc and PhD in experimental nuclear physics from Kyushu University. His research interests include Monte Carlo simulation of ionizing radiation and its application to medicine and biology.

Dimitris Emfietzoglou is an assistant professor in the Medical Physics Laboratory of the University of Ioannina Medical School. He graduated from the Department of Physics of the University of Athens and received his MSc and PhD in radiation science from Georgetown University in Washington, DC. His research interests include theoretical and computational aspects of the interaction of ionizing radiation with biomaterials  and nanostructures, and Monte Carlo particle transport simulation.

Section I

1

Introduction

Figure 1.1 is simulated for the passage of a radiation track through the human cell nucleus using Monte Carlo track structure technology. The track is crossing some of the 23 pairs of simulated chromosome volumes. It is a simple picture but tells a thousand-word story. In this book we tell you about the interaction of radiation tracks in the biological medium.

Interaction of radiation with biological matter in the human cell nucleus induces genomic instability, DNA damage, mutation, or cell death. What are the consequences to the cell and the organism in which it resides when perceived as irreducible systems?

Radiation is a double-edged sword. It is a harmful agent to organic and inanimate matter; on the other hand, it is a tool vastly used as a drug to eradicate cancerous cells. Radiation interacts with matter by depositing energy in the form of ionizations and excitations. These physical events leave a history behind that is known as track. Tracks come in many forms—long and short, thin and fat, sparse and dense. All these attributes describe the nature of the particle (or the radiation), whether it is a photon, electron, or heavy ion. Figure 1.1 shows a track segment of 1 MeV/u α-particles crossing a human cell nucleus. It shows interactions of the particle with the atoms in the medium. These interactions, shown as dots, seem to be very close to each other. Intuitions and perception could be very misleading in this case. The diameter of this particular cell is about 10 μm. How close are these interactions? Does it matter to the cell if these interactions are densely close or sparsely spaced? Why are they closely or sparsely spaced? The track shows branches coming off the center line. Some particles produce short but some very long branches. Is there significance to these? Some particles stop in matter completely, leaving behind a short track; some travel through a thick tissue before coming to rest. What is significant of such radiations and how useful are they in industrial use or in medical applications?

This book is about radiation tracks. We explore the physics of radiation tracks, share with you our knowledge gained over the past 30 years in this topic, and show you how radiation tracks can be simulated in a computer experiment. Interaction of radiation with matter is a beautiful subject, and radiation track is the story of a history to tell. This book takes you from the classical physics of track description to the modern aspects of condensed physics in matter. We start the learning with the essential preliminary knowledge and progress from there. There are many questions, exercises, and problems to reinforce the learning and the experience. The authors of

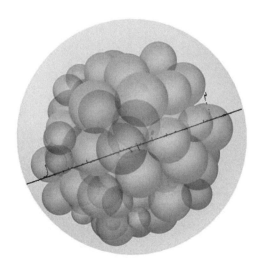

FIGURE 1.1 (SEE COLOR INSERT)
A 1 MeV/u α-particle crossing a cell nucleus.

this book are passionate about their science, and they want to share the knowledge with the young scientists and the students.

Discovery of radiation started with the work of scientific world luminaries Michael Faraday, James Clark Maxwell, Einstein, J.J. Thompson, Madame Curie, and others who established the nature of light and electromagnetic radiation. Since then much progress and insight have been gained in our scientific knowledge. Today, biology presents us as the most complex scientific discipline with many unsolved fundamental questions. A typical human cell nucleus, nearly 10 μm in diameter, contains over a billion DNA molecules. The compact DNA when unwound will stretch to about 180 cm. By current estimates, the human genome contains nearly 100,000 proteins and some 20,000 genes. This is the most complex system in the living environment. What type of physics and mathematics do we need to describe such a complex system that is neither linear nor thermodynamically a closed system? The subject of this book is far from the discussion of biological complexities, but we hope through the discussion of interaction of radiation with matter we open that frontier to greater scrutiny. This book describes interactions of ionizing radiation with matter in its biologically simplest form—water—and develops descriptive models for such a system.

We now scrutinize Figure 1.1 in more detail at the microscopic level. Figure 1.2 brings the view closer to examine the track for a 200 nm length. In this segment of the track there are 507 primary ion (red dots) and 2,073 secondary electron interactions (black dots). As this track was generated in track segment mode, elastic interactions were not considered. Figure 1.3 demonstrates a number of tracks generated in "track segment" and "full slowing down" modes. These include electron, proton, and carbon ions.

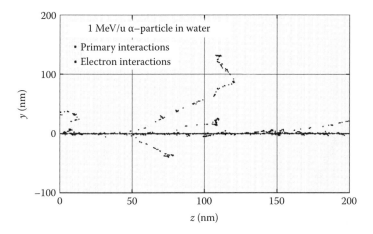

FIGURE 1.2 (SEE COLOR INSERT)
A short segment of the 1 MeV/u α-particle as in Figure 1.1.

Figure 1.4 provides a more advanced application of Monte Carlo track structure codes to construct depth-dose curves for ^{60}Co-γ photons and 200 MeV protons, typical radiations used in therapy. Table 1.1 gives some of the relevant numbers in the Bragg peak area, and in Table 1.2 some specific details of tracks illustrated in Figure 1.3 are listed.

1.1 Radiation Transport Codes

Tracks can be presented in many different ways. One of the earliest models of radiation track description is the linear energy transfer (LET). The concept of LET was originally introduced by Zirkle (1952), adopted from earlier notions of linear energy association (Zirkle 1940) and mean linear ion density (Gray 1947). Cormack and Johns (1952) introduced a more rigorous definition, but it did not separate the contributions of secondary electrons and the stochastic variations of energy loss along the track of the charged particles. Calculations of Burch and Bird (1956) and Burch (1957a,b) explicitly considered the contributions of secondary electrons generated by the interactions of the primary particle with atomic electrons along its trajectory. Burch's calculations were based on the particular criterion of 100 eV energy transfer, which was suggested by Gray (1947) as the dividing line between a primary and a secondary electron track. The secondary electrons resulting from interactions with a single energy transfer of >100 eV were considered to constitute a separate track. Energy transfers of ≤100 eV were considered to be deposited locally by the primary track. In radiation chemistry the chemical

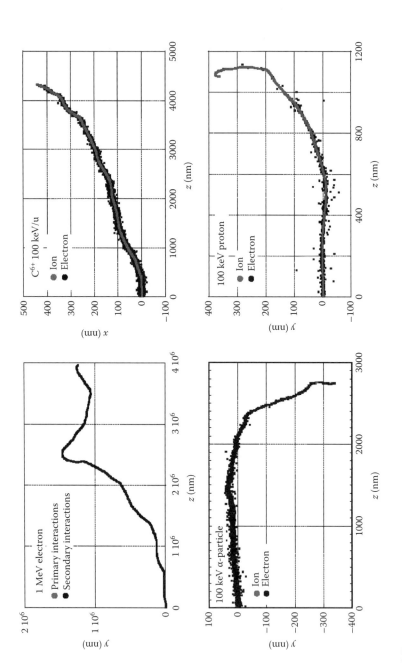

FIGURE 1.3

Four examples of Monte Carlo tracks generated by the code KURBUC for 1 MeV electron, full slowing down of a C^{6+} carbon ion, and a track of 100 keV proton. Tracks show points of energy depositions (ionizations and excitations) for the primary ion and the secondary electrons. All tracks were generated in water.

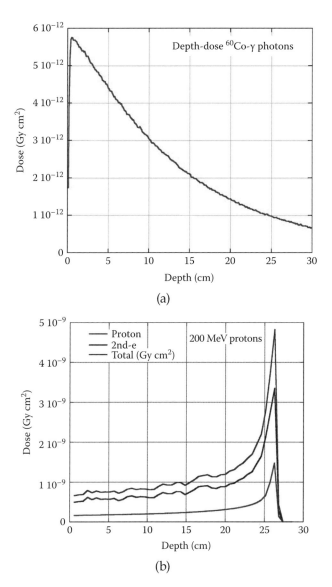

FIGURE 1.4 (SEE COLOR INSERT)
Depth-dose for ^{60}Co-γ photons and protons.

yield or number of molecules formed by reactions of energetic electrons is commonly expressed per 100 eV energy deposited by the track (Magee and Chatterjee 1987, Paretzke 1987).

A one-dimensional deterministic description of ion transport based on the solution of the Boltzmann transport equation has been developed at NASA Langley Research Center (Wilson et al. 1991). The development of these codes

TABLE 1.1

Number of Tracks Per Call

Radiation	Depth at	Dose (Gy cm²)	Fluence for 1 cGy/cm²	Average No. of Tracks per Cell
^{60}Co-γ rays	Bragg peak	$9e^{-13}$	1.72	4,444
200 MeV protons	Bragg peak	$4.7e^{-9}$	2.13	8.5

dates back to pre-NASA history and the work on medical problems of high-altitude travel by pilots. Many decades later a greater concern and awareness of the radiation problems and radiation risk in manned space flights led to the development of computer codes at the Langley Research Center (Wilson et al. 1989, 1991, 1995). These codes (HZETRN, BRYNTRN) are used to predict the propagation and interactions of charged particles, including the transport of high-energy ions up to iron or higher charged ions.

1.1.1 Amorphous Track Codes

The Katz track structure model was first published by Butts and Katz (1967); reviews were done by Katz (2003) and Katz et al. (1993), as well as extended versions of the model (Cucinotta et al. 1999; Wilson et al. 1993) and a modified version of the model (Scholz and Kraft 1992, 2004). The latter model has been used in carbon ion therapy planning at GSI (Kraft et al. 1999). The model of Katz was the first to introduce the concept of the lateral extension or radial distribution of dose around the track of ionizing radiation. The model does not consider the track structure of individual interactions but the radial dose profile represented as a homogenous dose distribution in the irradiated sample. The radial dose profile is determined by the ratio of charge to the velocity of the particle.

1.1.2 Condensed History Monte Carlo (CHMC) Codes

Demand for high-level accuracy in dosimetry, for example, absolute dose distributions in patients in radiotherapy and shielding problems, has led to the generation of general purpose Monte Carlo transport codes for energetic electrons and photons, and more recently for stripped ions, for arbitrary geometries. To date, there are a number of these general purpose codes, based on CHMC or other methods, listed in Table 1.3. The most widely used among these codes that are in public domain include EGS4 (Nelson et al. 1985) for the simulation of photons and electrons; MCNP (Goorley et al. 2003) and MCNPx (Hendricks et al. 2008) for simulation of neutrons and light ions; PHITS (Niita et al. 2010), a general purpose code for simulation of heavy ions up to 200GeV; FLUKA (Fasso et al. 2005) for all heavy ion track simulation; GEANT4 (Agnostelli et al. 2003); the Monte Carlo track structure code

TABLE 1.2

Number and types of events in track

	Electrons 1 MeV FSD	Proton 100 keV FSD	α-Particle 100 keV/u FSD	Proton 1 MeV 1 μm TS	Proton 1 MeV FSD	C⁶⁺ 100 keV/u FSD
Total number of interactions	107,471	13,846	54,962	2,845	112,505	127,729
Ionization by primary particle	15,034(14%)	1569(11%)	7,478(14%)	356(13%)	13,895(12%)	14,106(11%)
Excitation by primary particle	6,335(5.9%)	1,121(8.1%)	2,656(4.8%)	160(5.6%)	6,585(5.9%)	5,239(4.1%)
Electron capture	0(0.0%)	738(5.3%)	1,311(2.4%)	0(0.0%)	862(0.8%)	5,412(4.2%)
Electron loss	0(0.0%)	737(5.3%)	1,295(2.4%)	0(0.0%)	861(0.8%)	5,406(4.2%)
Elastic (ions)	5,585(5.2%)	3204(23%)	13,344(24%)	306(11%)	13,094(12%)	8,659(6.8%)
Ionization by secondary electrons	18,846(18%)	933(6.7%)	4,681(8.5%)	497(18%)	17,814(16%)	18,748(15%)
Excitation by secondary electrons	27,653(26%)	2424(18%)	10,436(19%)	719(25%)	28,121(25%)	32,516(26%)
Subexcitation electrons	34,018(32%)	3120(23%)	13,004(24%)	807(28%)	31,272(28%)	37,641(30%)

Note: FSD = full slowing down track; TS = track segment (1 μm).

TABLE 1.3

Partial List of General Purpose Monte Carlo Codes

Code	Particle	Energy Range	Reference
ETRAN	e⁻, phot	10 keV–1 GeV	Berger and Seltzer 1973
EGS4	e⁻, phot	10 keV–1 GeV	Nelson et al. 1985
FLUKA	p, n, Meson	1 keV–GeV	Fasso et al. 2005
GEANT4	p, n, Meson	250 eV–GeV	Agostinelli et al. 2003
MCEP	e⁻, phot	1 keV–30 MeV	Uehara 1986
MCNP5	n, phot, e⁻	See ref.	Goorley et al. 2003
MCNPX	n, light ions	See ref.	Hendricks et al. 2005
PENPENELOPE	e⁻, e+	100 eV–1 GeV	Salvat et al. 2008
PHITS	HZE	MeV–GeV	Niita et al. 2010
PEREGRINE	e⁻, phot	Therapy beams	Hartmann Siantar et al. 1998
PTRAN	Protons	<250 MeV	Berger 1993
SRIM	All ions	keV–2 GeV/u	Ziegler et al. 2003
SHIELD-HIT	$1 < Z < 10$	$1\ MeVu^{-1}$–$1\ TeVu^{-1}$	Sobolovsky 2010

PENELOPE (Salvat et al. 2008) for performing simulation of coupled electron-photon transport in arbitrary materials and complex quadric geometries in the range of 100 eV to 1 GeV; and SHIELD-HIT (Sobolovsky 2010) for the simulation of heavy ions and complex geometries.

1.1.3 3D and 4D Monte Carlo Track Structure Codes

The full Monte Carlo track structure codes provide the distribution of coordinates of all interactions of the charged particle in space (three-dimensional (3D) codes) and with time (four-dimensional (4D) codes). The 3D codes provide the distribution of physical events (ionization, excitations, and elastic scatterings). This is usually called the *physical track*. In a biological medium, physical events lead to the generation of radicals. The 4D track or the *chemical track* describes the distribution of chemical events in the medium with time. The timeframe for the physical track is fixed from the time of the first interaction at 10^{-15} to 10^{-13} s.

Table 1.4 provides a list of some currently published track structure codes used at various research centers around the world. The majority of the codes use water as the medium for simulation. Most experimental data are obtained in water vapor, as water in any phase is not a tractable medium as a target in experiments. In Section III we discuss in some detail the inelastic interaction of charged particles in condensed media within the context of the dielectric theory, which represents the state of the art for track structure simulations in liquid water. In the codes for liquid water a mixture of data derived from liquid and vapor targets has been employed, as currently there are no direct experimental data for excitation and ionization cross sections for liquid water.

TABLE 1.4

Partial List of Monte Carlo Track Structure Code[a]

Code	Particle	Energy Range	Reference
CPA100	e⁻	10 eV–100 keV	Terrissol and Beaudre 1990
ETRACK	e⁻, p, α	10 eV–10 keV e⁻	Ito 1987
KURBUC CODES	e⁻	10 eV–10 MeV	Uehara et al. 1993
	p	1 keV–300 MeV	Uehara et al. 2001, Liamsuwan et al. 2011
	α	1 keV/u–2 MeV/u	Uehara and Nikjoo 2002
	C ions	1 keV/u–10 MeV/u	Liamsuwan et al. 2012
LEEPS	e⁻, e⁺	≥10 eV–100 keV	Fernandez-Varea et al. 1996
MC4	e⁻, ions	≥10 eV e⁻, ions ≥ 0.3 MeV/u	Emfietzoglou et al. 2003
NOTRE DAME	e⁻, ions	≥10 eV e⁻, ions ≥ 0.3 MeV/u	Pimblott et al. 1990
PARTRAC[a]	e⁻, ions	≥10 eVe⁻, ions ≥ 0.3 MeV/u	Friedland et al. 2003
PITS99	e⁻, ions	≥ 10 eV e⁻, ions ≥ 0.3 MeV/u	Wilson and Nikjoo 1999
SHERBROOKE[a]	e⁻, ions	≥10 eV e⁻, ions > 0.3 MeV/u	Cobut et al. 2004

[a] For more details, see Nikjoo et al. (2006).

QUESTIONS

1. Give five examples of the way a radiation track (for example, an α-particle track) can be characterized.
2. Given the depth-dose values in Table 1.1, calculate fluence values for ⁶⁰Co-γ photons and 200 MeV protons. For the cell assume 20 μm diameter/side, spherical or cubic.

References

Agostinelli A, Allison J, Amako K et al. 2003. GEANT4; A SIMULATION TOOLKIT. *Nuclear Instrum. Methods Phys. Res.* A 506. 250–303. http://geant4.cern.ch/

Berger MJ, Seltzer SM. 1973. ETRAN, Monte Carlo Code System for electron and photon transport through extended media. ORNL Documentation for RISC Computer code package CCC-107.

Berger MJ. 1993. Proton Monte Carlo transport program PTRAN. Report NISTIR-5113. National Institute of Standards and Technology (NIST). http://www.nea.fr/abbs/html/ccc-0618.html

Burch PRJ, Bird PM. 1956. Linear energy transfer calculations allowing for delta tracks. *Progress in Radiobiology.* Birmingham: The Keynoch Press, pp. 161–177.

Burch PRJ. 1957a. Some physical aspects of relative biological efficiency *Br. J. Radiology.* 30: 524–529.

Burch PRJ. 1957b. Calculations of energy dissipation characteristics in water for various radiations. *Radiation Research.* 6: 289–301.

Butts JJ, Katz R. 1967. Theory of RBE for heavy ion bombardment of dry enzymes and viruses. *Radiat. Res.* 30: 855–871.

Cobut V, Cirioni L, Patau JP. 2004. Accurate transport simulation of electron tracks in the energy range 1keV-4MeV. *Nucl. Instrum. Methods. B.* 215: 57–68.

Cormack DV, Johns HE. 1952. Electron energies and ion densities in water irradiated with 200 keV, 1 MeV and 25 MeV radiation. *Br. J. Radiol.* 25: 369.

Cucinotta FA, Nikjoo H, Goodhead DT. 1999. Applications of amorphous track models in radiation biology. *Radiation and Environmental Biophysics.* 38: 81–92.

Emfietzoglou D, Karava K, Papamichael G, Moscovitch M. 2003. Monte Carlo simulation of the enrgy loss of low-energy electron in liquid water. *Phys. Med. Biol.* 48: 2355–2371.

Fasso A, Ferrari A, Ranft J, Sala P. 2005. Current version of FLUKA 2011.2.8. http://fluka.org/

Fernandez-Varea JM, Liljequist D, Csillag S, Rdty R, Salvat F. 1996. Monter Carlo simulation of 0.1-100 keV electron and positron transport in solids using optical data and partial wave methods. *Nucl. Instrum. Methods. B.* 108: 35–50.

Friedland W, Bernhard P, Jacob P, Paretzke HG, Dingfelder M. 2003. Simulation of DNA damage aftrer proton irradiation. *Radiat. Res.* 159: 401–410.

Goorley T, Brown F, Cox LJ. 2003. MCNP5™ Improvements for Windows PCs. Proc. M&C 2003: A Century in Review, A Century Anew. Gatlinburg, Tennessee. (LA-UR-02-7162) http://mcnp-green.lanl.gov/

Gray LH. 1947. The distribution of the ions resulting from the irradiation of living cells. *Br. J. Radiol.* (Suppl 1): 7.

Hartmann-Siantar CL, Moses EI. 1998. The PEREGRINE programmer: using physics and computer simulation to improve radiation therapy for cancer. *Eur. J. Phys.* 19: 513–521. https://www.llnl.gov/str/Moses.html

Hendricks JS, McKinney GW, Trellue HR, et al. 2008. MCNPx. Version 2.6.0. (LA-UR-08-2216.pdf) http://mcnpx.lanl.gov/

Ito A, 1987. Calculation of double strand break probability of DNA for low LET radiations based on track structure analysis. Nuclear and atomic data for radiotherapy and related radiobiology. International Atomic Energy Agency, IAEA, Vienna.

Katz R, Cucinotta FA, Wilson JW, Shinn JL, Ngo DM. 1993. A model of cell damage in space flight. In: Swenberg CE, Horneck G, Stassinopoulos EG (Eds.). *Biological Effects and Physics of Solar and Galactic Cosmic Radiation.* Part A. Plenum Press, New York pp. 235–268.

Katz R. 2003. The parameter-free track structure model of Scholz and Kraft for heavy-ion cross sections. *Radiat. Res.* 160: 724–728.

Kraft G, Scholz M, Bechthold U. 1999. Tumour therapy and track structure *Radiat. Environ. Biophys.* 38: 229–237.

Liamsuwan T, Nikjoo H. 2012. An energy-loss model for low- and intermediate-energy carbon projectiles in water. *Int. J. Radiat. Biol.* 88: 45–49.

Liamsuwan T, Uehara S, Emfietzoglou D, Nikjoo H. 2011. Physical and biophysical properties of proton tracks of energies 1 keV to 300 MeV in water. *Int. J. Radiat Biol.* 87: 141–160.

Magee JL, Chatterjee A. 1987. Track reactions of radiation chemistry in: Kinetics of nonhomogeneous processes. (Ed) GR Freeman, John Wiley and Sons, New York.

Nelson WR, Hirayama H, Rogers DWO. 1985. The EGS4 code system (SLAC report 265). http://rcwww.kek.jp/research/egs/

Niita K, Matsuda N, Iwamoto Y, Iwase H, Sato T, Nakashima H, Sakamoto Y, Sihver L. 2010. PHITS: Particle and Heavy Ion Transport code System, Version 2.23. JAEA-Data/Code 2010-022 (2010) http://phits.jaea.go.jp/

Nikjoo H, Uehara S, Emfietzoglou D, Cucinotta FA. 2006. Track-structure codes in radiation research. *Radiat. Measure.* 41: 1052–1074.

Paretzke HG. 1987. Radiation track structure theory, in: Kinetics of nonhomogeneous processes. (Ed) GR Freeman. John Wiley and Sons, New York.

Pimblott SM, La Verne JA, Mozumder A, Green NJB. 1990. Structure of electron tracks in water. 1. Distribution of energy deposition events. *J. Phys. Chem.* 94: 488–495.

Salvat F, Fernandez-Varea JM, Sempau J. 2008. PENELOPE-2008: A code system for Monte Carlo simulation of electron and photon transport. Issy-les-Moulineaux: OECD NuclearEnergy Agency. http://www.nea.fr/lists/penelope.html

Scholz M, Kraft G. 1992. A parameter-free track structure model for heavy ion action cross sections. In: *Biophysical modeling of radiation effects.* (Eds) Chadwick KH, Moschini G, Varma MN. EUR pub no. 13848 EN. Scientific and Technical Communication Unit, Commission of the EC, Directorate General Telecommunications, Information Industries and Innovation, Luxembourg, 185–192.

Scholz M, Kraft G. 2004. The physical and radiobiological basis of the local effect model: a response to the commentary by R. Katz. *Radiat. Res.* 161: 612–620.

Sobololevsky N. 2010. SHIELD-Hit: multipurpose hadron transport code. http://www.inr.ru/shield/

Terrissol M, Beaudre BA, 1990. Simulation of space and time evolution of radiolytic species induced by electrons in water. *Radiat. Protec. Dosim.* 31: 175–177.

Uehara S, Nikjoo H, Goodhead DT. 1993. Cross-sections for water vapour for Monte Carlo electron track structure code from 10 eV to 10 MeV region. *Phys. Med. Biol.* 38: 1841–1858.

Uehara S, Nikjoo H. 2002. Monte Carlo track structure code for low-energy alpha particles in water. *J. Phys. Chem.* 106: 11051–11063.

Uehara S, Toburen LH, Nikjoo H. 2001. Development of a Monte Carlo track structure code for low-energy protons in water. *Int. J. Radiat. Biol.* 77: 138–154.

Uehara S. 1986. The development of a Monte Carlo code simulating electron-photon showers and its evaluation by various transport benchmarks. *Nucl. Instrum. Methods B.* 14: 559–570.

Wilson JW, Badavi FF, Cucinotta FA, Shinn JL, Badhwar GD, Silberberg R, Tsao CH, Townsend LW, Tripathi RK. 1995. HZETRN: description of a free-space ion and nucleon transport and shielding computer program. NASA Technical paper 3495.

Wilson JW, Cucinotta FA, Shinn JL. 1993. Cell kinetics and track structure. In: Swenberg CE, Horneck G, Stassinopoulos EG (Eds). *Biological effects and physics of solar and galactic cosmic radiation.* Part A. Plenum Press, New York, 295–338.

Wilson JW, Townsend LW, Chen SY, Buck WW, Khan F, Cucinotta FA. 1989. BRYNTRN: baryon transport computer code. NASA TM 4037 and NASA TP 288.

Wilson JW, Townsend LW, Schimmerling W, Khandelwal GS, Khan F, Nealy JE, Cucinotta FA, Simonsen LC, Norbury JW. 1991. Transport methods and interactions for space radiations. RP1257, NASA Washington DC.

Wilson WE, Nikjoo H. 1999. A Monte Carlo code for positive ion track simulation. *Radiat. Environ. Biophys.* 38: 97–104.

Ziegler JF, Biersack JP, Littmark U. 2003. The stopping and ranges of ions and solids. Pergammon Press, New York. http://www.srim.org

Zirkle RE, Marchbank DF, Kuck KD. 1952. Exponential and sigmoid survival curves resulting from alpha and x-irradiation of Aspergillas spores. *J. Cell Comp. Physio.* 39 (Suppl. 1): 75–85.

Zirkle RE. 1940. The radiobiological importance of the energy distribution along ionization tracks. *J. Cell. Comp. Physiol.* 16: 221–235.

2

Basic Knowledge of Radiation

2.1 Definitions of Radiation

Radiation originally meant α-rays, β-rays, and γ-rays emitted from natural radioactive isotopes. At present, elementary particles, nuclei, electrons, and photons, moving with speeds comparable to or higher than the rays from radioactive isotopes, are called radiation. The reason for using the term *rays* is that the path of particles has direction. When radiation passes through matter, it has the capability of direct or indirect ionization of atoms and molecules. Hence we use the term *ionizing radiation*.

Radiation is separated into charged particles and noncharged particles. The former is called direct ionizing radiation because that directly ionizes atoms and molecules with its electric charge. X-rays, γ-rays, and neutrons belong to the latter. Since those do not have the electric charge, they cannot ionize using the electric force directly. However, particles interact with matter and generate secondary charged particles. These particles are called indirect ionizing radiation because secondary charged particles ionize atoms and molecules. Table 2.1 shows a list of radiation relevant to medical and clinical fields.

Radiation is defined as the particles with sufficient energy to cause ionization. If an accelerator is used, all ions are qualified as radiation. Therefore, all elements from hydrogen to uranium can be called radiation. In practice, heavy ions up to neon are utilized in the clinical field. The generation source of radiation is the accelerator or radionuclides except cosmic rays. The radiation sources are as follows:

X-rays: X-ray tube, linear accelerator, synchrotron radiation.

γ-Rays: Radioisotopes.

Electrons: Linear accelerator, betatron, microtron.

β-Rays: Radioisotopes.

Heavy ions: Linear accelerator, cyclotron, synchrotron.

Neutrons: Cyclotron, nuclear reactor, Cf-254, Cf-252.

TABLE 2.1

Ionizing Radiation

Name	Symbol	Electric Charge (e)	Mass (m_e)	Mean Life (s)	Production Method	Remarks
γ-Ray	γ	0	0		Radioisotope (RI)	Single energy
X-ray	X	0	0		Accelerator	Continuous energy
Neutrino	ν	0	~0		Radioisotope	β-decay
Electron (β⁻ ray)	e^-, $β^-$	−1	1		Accelerator, RI	β-ray is continuous energy
Positron (β⁺ ray)	e^+, $β^+$	+1	1		Accelerator, RI	Annihilation quanta
Proton	p	+1	1,836		Accelerator	
Neutron	n	0	1,839	1.1×10^3	Accelerator, Nuclear reactor	
Deuteron	d	+1	3,670		Accelerator	
Triton	t	+1	5,479	10^9	Accelerator	
α-Particle	α	+2	7,249		Accelerator, RI	
Muon	$μ^±$	±1	207	2.15×10^{-6}	High energy nuclear reaction	
Pion (charged)	$π^±$	±1	273	2.65×10^{-8}	High-energy nuclear reaction	
Pion (neutral)	$π^0$	0	264		High energy nuclear reaction	
Fission fragment (light)		~36	~96 m_p		Nuclear fission	
Fission fragment (heavy)		~56	~140 m_p		Nuclear fission	

2.2 Electron Volt

When an electric charge q accelerates through an electrostatic potential difference of V, the mechanical work W where the electric field acts on the charge is given by

$$W = qV \tag{2.1}$$

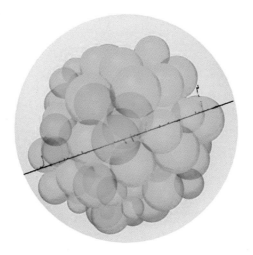

FIGURE 1.1
A 1 MeV/u α-particle crossing a cell nucleus.

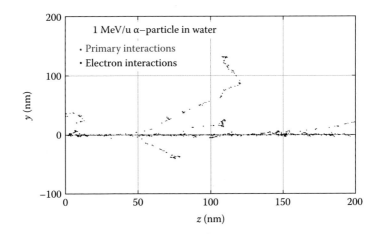

FIGURE 1.2
A short segment of the 1 McV/u α-particle as in Figure 1.1.

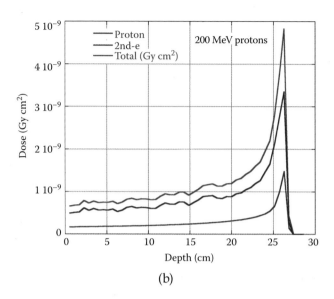

FIGURE 1.4
Depth-dose for ^{60}Co-γ photons and protons.

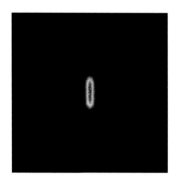

SOLUTION 11.7

This quantity is independent of the distance between both electrodes. The electric charge gains the kinetic energy corresponding to W. Since electric charge in nature cannot be divided infinitely, the minimum unit is called the elementary electric charge. An electron (e^-) has a negative elementary electric charge, $-e$.

$$e = 1.602177 \times 10^{-19} \text{C} \tag{2.2}$$

The electron rest mass m_e is

$$m_e = 9.10939 \times 10^{-31} \text{kg} \tag{2.3}$$

The kinetic energy that an electron obtains in motion between the potential of 1 V is defined by 1 eV (electron volt). The relationship between energy units eV and J using Equation (2.1) becomes

$$1 \text{eV} = 1.602177 \times 10^{-19} \text{J} \tag{2.4}$$

EXAMPLE 2.1

If an x-ray tube works with an electrical potential of 120 kV, what is the electron potential energy on the cathode surface?

SOLUTION 2.1

$$W = qV = (-1.6 \cdot 10^{-19} \text{ C})(-120 \text{ kV}) = 1.92 \cdot 10^{-14} \text{ J}$$

2.3 Special Theory of Relativity

The special theory of relativity was proposed in 1905 by Albert Einstein. It was constructed based on two fundamental principles:

1. The physical basic laws are not affected for all inertial frames (the principle of relativity).
2. Light in vacuum propagates with the speed c (a fixed constant) in terms of any system of inertial coordinates, regardless of the state of motion of the light source (the principle of invariant light speed).

Once we accept the principle of invariant light speed, a significant change on the concept of time and space must be made. For example, it is assumed that a light is emitted from the jet plane with a supersonic speed of v to the same direction as the plane. The light speed should be $c + v$ if the generally accepted theory is applied. However, the new principle gives the solution of c. In order to realize this principle, a new transformation of the coordinate system is required to take into account that time is needed. According to the special theory of relativity, the mass of body is revised. If a body with the rest mass m_0 moves with speed v, the mass of body m is represented by

$$m = \frac{m_0}{\sqrt{1 - v^2/c^2}} \tag{2.5}$$

In addition, total energy of the body and its mass are related by the equation

$$E = \frac{m_0 c^2}{\sqrt{1 - v^2/c^2}} = mc^2 \tag{2.6}$$

in which the equation $E = mc^2$ shows the mass m and energy E are equivalent. If the body rests, i.e., $v = 0$, E becomes $m_0 c^2$, which is called the rest energy. The relativistic momentum $p = mv$ is represented by

$$p = \frac{m_0 v}{\sqrt{1 - v^2/c^2}} \tag{2.7}$$

Combining Equations (2.6) and (2.7),

$$E^2 = m_0^2 c^4 + p^2 c^2 \tag{2.8}$$

is obtained.

For photons the rest mass is 0. Inserting $m_0 = 0$ into Equation (2.8),

$$E = pc \tag{2.9}$$

This equation means the energy and the momentum have nonzero finite value even for the particle of $m_0 = 0$.

EXAMPLE 2.2

What is the energy equivalent of the electron and proton with masses $9.109384 \cdot 10^{-31}$ kg and $1.672623 \cdot 10^{-27}$ kg ?

SOLUTION 2.2

$E = mc^2$

$E_e = (9.109384 \cdot 10^{-31} \text{ kg})(2.998 \cdot 10^8 \text{ m/s})^2 = (8.1875 \cdot 10^{-14} \text{ J})\left(\dfrac{1 \text{ MeV}}{1.602 \cdot 10^{-13} \text{ J}} \right)$

$= 0.511 \text{ MeV}$

$E_p = (1.672623 \cdot 10^{-27} \text{ kg})(2.998 \cdot 10^8 \text{ m/s})^2 = (1.5033 \cdot 10^{-10} \text{ J})\left(\dfrac{1 \text{ MeV}}{1.602 \cdot 10^{-13} \text{ J}} \right)$

$= 938 \text{ MeV}$

EXAMPLE 2.3

Compare the relativistic and nonrelativistic kinetic energy of an electron traveling with a speed of 0.01, 0.1, 0.5, and 0.9 of the speed of light.

SOLUTION 2.3

The relativistic kinetic energy is

$$T_{rel} = m_0 c^2 \left(\frac{1}{\sqrt{1 - \frac{v^2}{c^2}}} - 1 \right).$$

The nonrelativistic kinetic energy is given by $T_{non-rel} = \frac{1}{2} m_0 v^2$.
m_0 is the rest mass of the electron ($m_0 = 9.10938 \cdot 10^{-31}$ kg), $c = 2.99792 \cdot 10^8$ m/s is the speed of light, and e is the elementary charge. Results are given in the following table:

Table Example 2.3

$\frac{v}{c}$	$\frac{1}{\sqrt{1-\frac{v^2}{c^2}}} - 1$	T_{rel} (J)	T_{rel} (eV) = $\frac{T_{rel} \text{(j)}}{e}$	$T_{non-rel}$ (J)	$T_{non-rel}$ (eV)= $\frac{T_{non-rel} \text{(j)}}{e}$	Deviation of $T_{non-rel}$ from T_{rel} (%)
0.01	5.0004E−05	4.0939E−18	2.5587E+01	4.0936E−18	2.5585E+01	0.0075
0.1	5.0378E−03	4.1245E−16	2.5778E+03	4.0936E−16	2.5585E+03	0.7506
0.5	1.5470E−01	1.2665E−14	7.9159E+04	1.0234E−14	6.3962E+04	19.1987
0.9	1.2942E+00	1.0595E−13	6.6221E+05	3.3158E−14	2.0724E+05	68.7055

2.4 Electromagnetic Wave and Photon

In classical electromagnetism, Maxwell's equations are a set of four equations that describe the properties of the electric and magnetic fields and relate them to their sources, charge density, and current density. The variation of the electric and magnetic field relates mutually. Even if the electric charge or the electric current varies, the electromagnetic field generated in the surrounding space cannot vary by a whole field in a moment. The field close to the current varies immediately; however, the variation will delay at the distant space. This means the variation of the electric and magnetic field propagates with a finite speed. This is the electromagnetic wave.

The propagation velocity of the electromagnetic wave, v, is given by

$$v = \frac{1}{\sqrt{\varepsilon_0 \mu_0}} \tag{2.10}$$

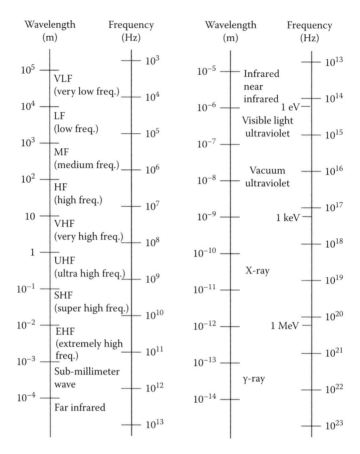

FIGURE 2.1
Name, wavelength, and frequency of electromagnetic wave.

where ε_0 and μ_0 are the permittivity of free space and permeability of free space, respectively. Putting v into c,

$$c = 2.998 \times 10^8 \quad \text{ms}^{-1} \tag{2.11}$$

is equal to the light speed. Therefore, the electric and magnetic field is the wave motion with the light velocity c inducing mutually. The electromagnetic wave is classified into various names, depending on its origin. Figure 2.1 shows the name, wavelength, and frequency of the electromagnetic wave.

On the other hand, Einstein constructed the photon theory in order to explain the photoelectric effect using Planck's idea of energy quantum. The light with the frequency v and the wavelength λ can be regarded with a particle having energy and momentum such as

$$E = h v, \quad p = h/\lambda = h v/c \tag{2.12}$$

where h is Planck's constant with the value of 6.626×10^{-34} Js, which is an important physical constant appearing in the law of microscopic nature. This particle is called a light quantum or a photon. The photon theory was completely recognized by a successful explanation for the Compton effect discovered by Compton in 1923. On the other hand, we cannot abandon the idea of electromagnetic wave in order to interpret the phenomena such as interference and diffraction of light. Therefore, we need to accept the wave-particle duality for the property of light. A unified theory describing both properties was constructed by quantum mechanics. A relationship between the photon energy E (keV) and the wavelength λ (nm) is derived from Equation (2.12).

$$\lambda = \frac{ch}{E} = \frac{1.240}{E} \tag{2.13}$$

EXAMPLE 2.4

Calculate the frequency and the wavelength of 1 MeV x-rays.

SOLUTION 2.4

$$v = \frac{E}{h} = \frac{(1\text{MeV})\left(\dfrac{1.602 \cdot 10^{-13}\ \text{J}}{1\text{MeV}}\right)}{6.626 \cdot 10^{-34}} = 2.417 \cdot 10^{20}\,\text{Hz}$$

$$\lambda = \frac{c}{v} = \frac{2.998 \cdot 10^{8}}{2.417 \cdot 10^{20}} = 1.24 \cdot 10^{-12}\,\text{m}$$

2.5 Interaction Cross Sections

Rutherford carried the classical mechanics calculations on scattering of α-particles by a positively charged nucleus. This calculation is a typical example of scattering of atomic or molecular collisions. That is very useful for understanding the concept of interaction cross sections. First, the concept of cross section is explained. The intensity of incident particles is assumed to be I, which gives the number of particles passing through a vertical unit area during a unit time. The direction of incident particle is changed by the collision with a target. That is called scattering. The probability of scattering into a given direction is represented by the differential cross section $d\sigma/d\Omega$. Its definition is as follows:

$$\frac{d\sigma}{d\Omega} = \frac{\text{number of particles scattered into the solid angle } d\Omega \text{ a unit time}}{\text{intensity of the incident particles}} \tag{2.14}$$

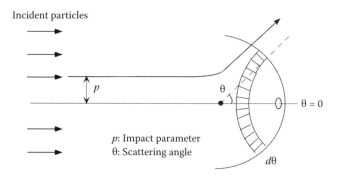

FIGURE 2.2
Scattering of the projectile by a central force.

As shown in Figure 2.2, the elementary solid angle is usually given by

$$d\Omega = 2\pi \sin\theta d\theta \tag{2.15}$$

because of the symmetrical property around the incident axis of particles. The term *cross section* arises because σ has the dimension of area.

The magnitude of scattering is determined by an impact parameter p. The number of particles scattered into the solid angle $d\Omega$ between θ and $\theta + d\theta$ should be equal to the number of incident particles with the corresponding impact parameter between p and $p + dp$. Therefore,

$$2\pi Ipdp = -2\pi I \frac{d\sigma}{d\Omega} \sin\theta d\theta \tag{2.16}$$

If p increases by dp, then the force acting on the particle becomes weaker and the scattering angle reduces by $d\theta$. This is the reason for the minus sign on the right-hand side. From Equation (2.16), $d\sigma/d\Omega$ becomes

$$\frac{d\sigma}{d\Omega} = -\frac{p}{\sin\theta} \frac{dp}{d\theta} \tag{2.17}$$

The differential cross section can be evaluated using this formula if the relationship between p and θ is obtained. The total cross section σ_t is provided by integrating over the full solid angle.

$$\sigma_t = \int_{4\pi} \frac{d\sigma}{d\Omega} d\Omega = 2\pi \int_0^\pi \frac{d\sigma}{d\Omega} \sin\theta d\theta \tag{2.18}$$

Next, Rutherford scattering is explained. The problem is the motion of two positive charges acting on each other by the electric force as inverse square of distance. Because the nucleus is heavier than the α-particle, the nucleus can be regarded as the center of the force fixed in the space. The α-particle approaches the nucleus from an infinite far distance and changes the direction and goes away to an infinite far distance; therefore, its orbit becomes a hyperbola. Here, we show just the solution, omitting the mathematical procedures. The differential cross section of the Rutherford scattering is represented by

$$\frac{d\sigma}{d\Omega} = \frac{1}{16}\left(\frac{zZe^2}{4\pi\varepsilon_0 E}\right)^2 \frac{1}{\sin^4\frac{\theta}{2}} \tag{2.19}$$

in which the scattering angle is θ, the nuclear charge is Z, the charge of the α-particle is z, the mass is m, and $E = mv^2/2$. Since this formula does not include the screening effect by orbital electrons, the total cross section σ_t calculated using Equation (2.18) diverges to infinity. If the screening effect is taken into account, it does not diverge.

EXAMPLE 2.5

In the Rutherford's scattering experiment, it was found that the fraction of α-particles scattered through an angle θ by a gold foil is $3.7\cdot10^{-7}$. The gold foil is $2.1\cdot10^{-5}$ cm thick. The scattered α-particles were counted on a screen of 1 mm² area placed at a distance of 1 cm from the scattering foil. What is the differential cross section of this collision at the angle θ?

SOLUTION 2.5

From Equation (2.14), the differential cross section can be written as

$$\frac{d\sigma}{d\Omega} = \left(\frac{N_s}{N_i}\right)\frac{\Delta S}{\Delta\Omega\cdot N_t}$$

where N_i and N_s are the number of incident and scattered α-particles, respectively; $\Delta\Omega$ is the solid angle through which the particles were scattered; and N_t is the number of scattering centers.

Assuming that the area ΔS of the gold foil was impinged by the beam of α-particles, this foil has the thickness d, the density ρ, and the atomic weight A. Thus, N_t is calculated using

$$N_t = \frac{\rho N_A d\Delta S}{A}$$

where N_A is the Avogadro's constant. The solid angle can be written as

$$\Delta\Omega = \frac{\Delta a}{r^2}$$

where Δa is the counting area placed at a distance r from the gold foil.

Finally, the differential cross section of this collision is given by

$$\frac{d\sigma}{d\Omega} = \left(\frac{N_s}{N_i}\right)\frac{Ar^2}{\rho N_A d\Delta a}$$

$$= (3.5\cdot10^{-7})\frac{(0.197 \text{ kg/mol})(10^{-2} \text{ m})^2}{(1.93\cdot10^4 \text{ kg/m}^3)\cdot(6.022\cdot10^{23} \text{ mol}^{-1})(2.1\cdot10^7 \text{ m})(10^{-6} \text{ m}^2)}$$

$$= 2.825\cdot10^{-27} \text{ m}^2 = 28.25 \text{ b}$$

2.6 Quantities and Units of Radiation

When we discuss radiation interaction with matter, we need to know the definition and units of radiation quantities. The present descriptions are based on the recommendations of ICRU Report 33 (1980) and Report 51 (1993).

2.6.1 Relevant to Radiation Fields

1. Particle number (N): The number of particles emitted, penetrated, or incident. The unit is 1.
2. Radiant energy (R): The energy of particles emitted, penetrated, or incident. (The rest energy is not included.) The unit is J.
3. Particle flux ($\dot{N}$):

$$\dot{N} = \frac{dN}{dt} \tag{2.20}$$

dN is the increment of particles within the time dt. The unit is s^{-1}.

4. Energy flux ($\dot{R}$):

$$\dot{R} = \frac{dR}{dt} \tag{2.21}$$

dR is the increment of radiant energy within the time dt. The unit is W.

5. Particle fluence (Φ):

$$\Phi = \frac{dN}{da} \tag{2.22}$$

Φ is the quotient of the particle number dN incident on the sphere divided by da, the sphere cross section. The unit is m^{-2}.

6. Energy fluence (Ψ):

$$\Psi = \frac{dR}{da} \qquad (2.23)$$

Ψ is the quotient of the radiant energy dR incident on the sphere divided by da, the sphere cross section. The unit is Jm^{-2}.

7. Particle fluence rate (ϕ):

$$\phi = \frac{d\Phi}{dt} = \frac{d^2N}{dadt} \qquad (2.24)$$

$d\Phi$ is the increment of particle fluence within the time dt. The unit is $m^{-2}s^{-1}$.

8. Energy fluence rate (ψ):

$$\psi = \frac{d\Psi}{dt} = \frac{d^2R}{dadt} \qquad (2.25)$$

$d\Psi$ is the increment of energy fluence within the time dt. The unit is $W\,m^{-2}s^{-1}$.

2.6.2 Relevant to Interactions

2.6.2.1 Cross Section (σ)

$$\sigma = \frac{P}{\Phi} \qquad (2.26)$$

P is the probability of interaction with a target (an atom or a molecule) when the particle fluence Φ hits. The unit is m^2. The unit frequently used is barn (10^{-28} m^2).

2.6.2.2 Mass Attenuation Coefficient (μ/ρ)

The mass attenuation coefficient for noncharged particles is given by

$$\frac{\mu}{\rho} = \frac{1}{\rho N}\frac{dN}{dl} \qquad (2.27)$$

N is the number of particles that entered the layer with the thickness dl and the density ρ. dN means the number of particles that interacted in this layer and varied their energies or directions. The unit is $m^2\,kg^{-1}$. For x-rays and γ rays,

$$\frac{\mu}{\rho} = \frac{\tau}{\rho} + \frac{\sigma_c}{\rho} + \frac{\sigma_{coh}}{\rho} + \frac{\kappa}{\rho} \qquad (2.28)$$

Each term on the right-hand side represents the mass attenuation coefficients for photoelectric effect, Compton effect, incoherent scattering, and electron pair creation, respectively.

2.6.2.3 Mass Energy Transfer Coefficient (μ_{tr}/ρ)

The mass energy transfer coefficient for noncharged particles is given by

$$\frac{\mu_{tr}}{\rho} = \frac{1}{\rho EN} \frac{dE_{tr}}{dl} \tag{2.29}$$

E is the total energy, except the rest energy, of noncharged particles that entered the layer with the thickness dl and the density ρ. dE_{tr} means the sum of kinetic energy generated in this layer. The unit is $m^2\,kg^{-1}$.

2.6.2.4 Mass Energy Absorption Coefficient (μ_{en}/ρ)

The mass energy absorption coefficient for noncharged particles is given by

$$\frac{\mu_{en}}{\rho} = \frac{\mu_{tr}}{\rho}(1-g) \tag{2.30}$$

where g is the fraction of the energy lost by a secondary charged particle in matter by bremsstrahlung with the unit $m^2\,kg^{-1}$. The difference between μ_{en}/ρ and μ_{tr}/ρ becomes significant when the kinetic energy of secondary charged particles is high and the atomic number of matter is large.

2.6.2.5 Total Mass Stopping Power (S/ρ)

The total mass stopping power of material for charged particles is given by

$$\frac{S}{\rho} = \frac{1}{\rho} \frac{dE}{dl} \tag{2.31}$$

dE is the energy loss of charged particles within the length dl in the matter with the density ρ. The unit is $Jm^2\,kg^{-1}$. S means the total linear stopping power. In the energy region where nuclear interaction can be neglected, the total mass stopping power is represented by

$$\frac{S}{\rho} = \frac{1}{\rho}\left(\frac{dE}{dl}\right)_{col} + \frac{1}{\rho}\left(\frac{dE}{dl}\right)_{rad} \tag{2.32}$$

where $(dE/dl)_{col} = S_{col}$ is the linear collision stopping power, and $(dE/dl)_{rad} = S_{rad}$ is the linear radiative stopping power.

2.6.2.6 LET (Linear Energy Transfer) or Restricted Linear Collision Stopping Power (L_Δ)

The linear energy transfer (LET) for charged particles in matter is given by

$$L_\Delta = \left(\frac{dE}{dl}\right)_\Delta \tag{2.33}$$

Δ represents the energy of secondary electrons generated by collision. The equation means the collision stopping power due to secondary electrons having energy below Δ. For example, L_{100} is called the LET for the cutoff energy of 100 eV. If all secondary electrons are taken into account, Δ becomes ∞. Therefore, L_∞ is equal to S_{col}. The unit is J m^{-1}. Usually the units are given in keV μm^{-1} or g/cm^2.

2.6.2.7 Radiation Chemical Yield (G)

The chemical yield of reaction products by absorption of radiation energy is called the G value.

$$G = \frac{n}{\bar{\varepsilon}} \tag{2.34}$$

in which $\bar{\varepsilon}$ is the absorbed energy and n is the produced numbers of an atom or a molecule. The unit is mol J^{-1}.

2.6.2.8 Average Energy per Ion Pair (W)

The average energy dissipated to form an ion pair in a gas is called W value. The unit is eV.

$$W = \frac{E}{\bar{N}} \tag{2.35}$$

$\bar{N}$ is the average number of produced ion pairs when the charged particle having the initial kinetic energy of E is completely dissipated in the gas.

2.6.3 Relevant to Doses

2.6.3.1 Energy Imparted (ε)

The radiation energy imparted to matter in a certain volume is given by

$$\varepsilon = R_{in} - R_{out} + \sum Q \tag{2.36}$$

in which, R_{in} = sum of radiant energy of all charged and noncharged particles entered in this volume (except the rest energy), R_{out} = sum of radiant energy of all charged and noncharged particles emitted from this volume (except the rest energy), and ΣQ = sum of the change of the rest energy caused by the nuclear transformation of nuclei and elementary particles in this volume.

The unit is J. The ε is a stochastic quantity, which is the fundamental quantity in microdosimetry.

2.6.3.2 Absorbed Dose (D)

The absorbed dose is defined by the division of mean energy deposition $d\bar{\varepsilon}$ (this is a nonstochastic quantity) by a given mass dm of matter.

$$D = \frac{d\bar{\varepsilon}}{dm} \tag{2.37}$$

The unit is J kg^{-1}. The specific name is gray (Gy): 1 Gy = 1 J kg^{-1}.

2.6.3.3 Absorbed Dose Rate ($\dot{D}$)

$$\dot{D} = \frac{dD}{dt} \tag{2.38}$$

dD is the increment of absorbed dose in the time dt. The specific unit name is 1 Gy s^{-1} = 1 J kg^{-1}.

2.6.3.4 Kerma (K)

This is the abbreviation for kinetic energy released in material. Kerma is defined by the division of the sum of initial kinetic energy of all charged particles released by noncharged particles by a given mass dm.

$$K = \frac{dE_{tr}}{dm} \tag{2.39}$$

The unit is J kg^{-1}. The specific name is gray (Gy); 1 Gy = 1 J kg^{-1}. Kerma is applicable only for noncharged particles. dE_{tr} is the initial energy of secondary charged particles; therefore, it includes energy loss not only by ionization and excitation but also to bremsstrahlung. In addition, energy of Auger electrons resulting from the photoelectric effect within the volume element is also included. For noncharged particles with energy E, a relationship between energy fluence Ψ and kerma K holds:

$$K = \Psi\left(\frac{\mu_{tr}}{\rho}\right) = \Phi\left[E\left(\frac{\mu_{tr}}{\rho}\right)\right] \tag{2.40}$$

where μ_{tr}/ρ is the mass energy transfer coefficient and $[E(\mu_{tr}/\rho)]$ is called kerma factor. If charged particle equilibrium holds in matter and bremsstrahlung loss can be ignored, absorbed dose D is equal to kerma K. For high-energy photons, charged particle equilibrium does not hold and K becomes slightly smaller than D.

2.6.3.5 Kerma Rate ($\dot{K}$)

$$\dot{K} = \frac{dK}{dt} \tag{2.41}$$

2.6.3.6 Exposure (X)

It is assumed that all electrons (both negative and positive) generated by incident photons completely stop in air with the mass dm. Electron-ion pairs are produced during electron slowing down. If the total charge of one pair is dQ, the exposure is defined by

$$X = \frac{dQ}{dm} \tag{2.42}$$

The unit is C kg^{-1}. The charge generated by bremsstrahlung due to secondary electrons is not included in dQ. It is difficult to measure the exposure in the energy range lower than a few keV or higher than a few MeV photons. An alternative definition of the exposure is

$$X = \Psi \frac{\mu_{en}}{\rho} \frac{e}{W} \tag{2.43}$$

2.6.3.7 Exposure Rate ($\dot{X}$)

$$\dot{X} = \frac{dX}{dt} \tag{2.44}$$

The unit is C kg^{-1} min^{-1} or C kg^{-1} h^{-1} usually.

2.6.4 Relevant to Radioactivities

2.6.4.1 Decay Constant (λ)

For a radionuclide at a specific energy state, the decay constant is the probability that the nucleus causes a spontaneous nuclear transition from that state. The unit is s^{-1}.

2.6.4.2 Activity (A)

The activity is the quantity of a radionuclide at a specific energy state at a time. It is assumed that the expectation value of the number of spontaneous nuclear transition from the state is dN within the time dt.

$$A = \frac{dN}{dt} \tag{2.45}$$

The unit is s^{-1}. The specific name is Bq. 1 Bq = 1 s^{-1}. Here, a specific energy state specifies the nuclear ground state. The activity A for a radionuclide at the specific energy state is equal to the product of λ and the number of nucleus N at the state.

$$A = \lambda N \tag{2.46}$$

EXAMPLE 2.6

The activity of a cobalt-60 source was measured to be 620 kBq on November 1, 1986. What is the activity of this source on October 31, 2011?

SOLUTION 2.6

The half-life, $T_{1/2}$, of Co-60 is 5.26 years. The decay constant λ is calculated using

$$\lambda = \frac{\ln 2}{T_{1/2}} = \frac{\ln 2}{5.26 \text{ y}} = 0.131777 \text{ y}^{-1}$$

From Equations (2.45) and (2.46), the activity A is given by

$$A = \lambda N = \frac{dN}{dt}$$

$$\Rightarrow N = N_0 e^{-\lambda t}$$

$$\Rightarrow A = A_0 e^{-\lambda t}$$

where the index 0 denotes the initial number of radioactive atoms, N, or activity of the source. Therefore, the activity of the source on October 31, 2011 (25 years later) is

$$A = 620 \text{ kBq} \cdot \exp(-0.13777 \text{ y}^{-1} \cdot 25 \text{ y}) = 22.995 \text{ kBq}$$

EXAMPLE 2.7

Calculate the exposure of 1 MeV photon in air, given the photon fluence of 10^9 cm^{-2}.

SOLUTION 2.7

The W value for air is $W_{air} = 34$ eV.

$$X = \phi E \left(\frac{\mu_{en}}{\rho} \right)_{air} \frac{e}{W_{air}}$$

$$= 10^9 \text{ cm}^{-2} \cdot 10^6 \text{ eV} \cdot \left(0.0278 \ \frac{\text{cm}^2}{\text{g}} \right) \left(\frac{1.6 \cdot 10^{-19} \text{ C}}{34 \text{ eV}} \right)$$

$$= 1.308 \cdot 10^{-7} \text{ C/g} = 1.308 \cdot 10^{-4} \text{ C/g}$$

EXAMPLE 2.8

Calculate the dose that the same photon beam would deposit in liquid water.

SOLUTION 2.8

$$D = \phi E \left(\frac{\mu_{en}}{\rho} \right)_{water}$$

$$= 10^9 \text{ cm}^{-2} \cdot 10^6 \text{ eV} \cdot \left(0.0309 \ \frac{\text{cm}^2}{\text{g}} \right)$$

$$= 3.09 \cdot 10^{13} \text{ eV/g} = 4.944 \cdot 10^{-3} \text{ Gy}$$

2.6.4.3 Air Kerma Rate Constant (Γ_δ)

A point source with the activity A is located at the distance l from the point of interest. At this point the air kerma rate constant for photons having energy greater than δ is given by

$$\Gamma_\delta = \frac{l^2 \dot{K}_\delta}{A} \tag{2.47}$$

where $\dot{K}_\delta$ is the air kerma rate. The unit is m^{2}Gy Bq^{-1}s^{-1}.

2.6.4.4 Exposure Rate Constant (Γ'_δ)

The activity A is located at the distance l from the point of interest. At this point the exposure rate constant for photons having an energy greater than δ is given by

$$\Gamma'_\delta = \frac{l^2}{A}\left(\frac{dX}{dt}\right)_\delta \tag{2.48}$$

The relevant photons include γ-rays, characteristic x-rays, and bremsstrahlungs.

2.6.5 Relevant to Radiation Protection

2.6.5.1 Dose Equivalent (H)

The effect of radiation action on biological material is different depending on the radiation type, even if for the same absorbed dose. In order to indicate the differences, the term *relative biological effectiveness* (RBE) has been used (Nikjoo and Lindborg 2010). The dose equivalent (H) is represented by

$$H = QD \tag{2.49}$$

in which D is the absorbed dose at a point in tissue and Q is the quality factor at that point. The unit of the dose equivalent is Sv if the unit of absorbed dose is Gy; 1 Sv = 1 Gy. Table 2.2 shows the summary of radiation quantities and units.

EXAMPLE 2.9

What is the kerma in air produced by the photon beam?

SOLUTION 2.9

$$K = \phi E\left(\frac{\mu_{tr}}{\rho}\right)_{air}$$

$$= 10^9 \text{ cm}^{-2} \cdot 10^6 \text{ eV} \cdot \left(0.028\frac{\text{cm}^2}{\text{g}}\right)$$

$$= 2.8 \cdot 10^{13} \text{ eV/g} = 4.48 \cdot 10^{-3} \text{ Gy}$$

TABLE 2.2

Radiation Quantities and Units

Name	Symbol	SI Unit	Specific Name of Unit	Special Unit
Particle number	N	1		
Radiant energy	R	J		
Particle flux	$\dot{N}$	s^{-1}		
Energy flux	$\dot{R}$	W		
Particle fluence	Φ	m^{-2}		
Energy fluence	Ψ	$J\,m^{-2}$		
Particle fluence rate	ϕ	$m^{-2}\,s^{-1}$		
Energy fluence rate	ψ	$W\,m^{-2}$		
Particle radiance	p	$m^{-2}\,s^{-1}\,sr^{-1}$		
Energy radiance	γ	$W\,m^{-2}\,sr^{-1}$		
Cross section	σ	m^2		b
Mass attenuation coefficient	μ/ρ	$m^2\,kg^{-1}$		
Mass energy transfer coefficient	μ_{tr}/ρ	$m^2\,kg^{-1}$		
Mass energy absorption coefficient	μ_{en}/ρ	$m^2\,kg^{-1}$		
Total mass stopping power	S/ρ	$J\,m^2\,kg^{-1}$		$eV\,m^2\,kg^{-1}$
Linear energy transfer	L_Δ	$J\,m^{-1}$		$eV\,m^{-1}$
Radiation chemical yield	$G(x)$	$mol\,J^{-1}$		
Mean energy per ion pair	W	J		eV
Energy imparted	ε	J		
Lineal energy	y	$J\,m^{-1}$		$eV\,m^{-1}$
Specific energy	z	$J\,kg^{-1}$	Gy	rad
Absorbed dose	D	$J\,kg^{-1}$	Gy	rad
Absorbed dose rate	$\dot{D}$	$J\,kg^{-1}\,s^{-1}$	$Gy\,s^{-1}$	$rad\,s^{-1}$
Kerma	K	$J\,kg^{-1}$	Gy	rad
Kerma rate	$\dot{K}$	$J\,kg^{-1}\,s^{-1}$	$Gy\,s^{-1}$	$rad\,s^{-1}$
Exposure	X	$C\,kg^{-1}$		R
Exposure rate	$\dot{X}$	$C\,kg^{-1}\,s^{-1}$		$R\,s^{-1}$
Decay constant	λ	s^{-1}		
Activity	A	s^{-1}	Bq	Ci
Air kerma rate constant	Γ_δ	$m^2\,J\,kg^{-1}$	$m^2\,Gy\,Bq^{-1}\,s^{-1}$	$m^2\,rad\,Ci^{-1}s^{-1}$
Dose equivalent	H	$J\,kg^{-1}$	Sv	rem
Dose equivalent rate	$\dot{H}$	$J\,kg^{-1}\,s^{-1}$	$Sv\,s^{-1}$	$rem\,s^{-1}$

2.7 Summary

1. Energetic elementary particles, nucleus, and photon moving in space and matter are called radiation.

2. When radiation travels in matter, particles with sufficient energy to ionize atoms and molecules are called ionizing radiation.

3. Charged particles are called directly ionizing radiation. Uncharged particles such as photons and neutrons are called indirectly ionizing radiation.

4. The energy of radiation is generally represented by electron volt (eV) units.

5. The theory of relativity changes the concepts of time, space, and mass. Therefore, energy and momentum in mechanics are represented in the manner of the relativistic theory.

6. The electromagnetic wave is derived from Maxwell's equations. X-rays and γ-rays are a kind of electromagnetic wave. Simultaneously, γ-rays and x-rays are also called photons because they have the characteristics of particles.

7. Energy and momentum of a photon having the frequency ν is $h\nu$ and $h\nu/c$, respectively.

8. Probability of reaction is represented by cross section. Energy spectrum or angular distribution is represented by the differential cross section.

QUESTIONS

1. The sequence of elements by atomic weight is Cl, K, Ar, and by atomic number using Mosely's law, Cl, Ar, K. Explain which is the correct sequence.
2. What is radiation pressure?
3. What can be seen on a screen if a flame is placed between two charged plates and a shadow of it is observed on the screen?

References

ICRU. 1980. *Radiation Quantities and Units*. ICRU Report 33.
ICRU. 1993. *Quantities and Units in Radiation Protection Dosimetry*. ICRU Report 51.
Nikjoo H, Linborg L. 2010. RBE of low energy electrons and protons. *Phys. Med. Biol.* 55: R65.

For Further Reading

Goldstein H. 1950. *Classical Mechanics*. Reading, MA: Addison-Wesley.
Johns HE, Cunningham JR. 1974. *The Physics of Radiology*, 3rd ed. Springfield, IL: Charles C. Thomas Publisher.
Turner JE. 1995. *Atoms, Radiation, and Radiation Protection*, 2nd ed. New York: John Wiley & Sons.

3

Atoms

3.1 Atomic Nature of Matter

At the beginning of the nineteenth century, the English chemist John Dalton discovered important laws that are the basis of modern chemistry. The first is the law of definite proportions. This law states that a chemical compound always contains exactly the same proportion of elements by mass. For example, oxygen makes up 8/9 of the mass of any sample of pure water, while hydrogen makes up the remaining 1/9 of the mass. In the case other than this ratio between oxygen and hydrogen, the elements do not make water, but whichever excess element remains. The second is the law of multiple proportions. The law states that when chemical elements combine, they do so in a ratio of small whole numbers. For example, a reaction of carbon of 12 g with oxygen of 32 g makes carbon dioxide (CO_2). Carbon monoxide (CO) is a combination of carbon of 12 g with oxygen of 16 g. In comparison with these two compounds, oxygen jumps by twice the amount, from 16 g to 32 g. Dalton's atomic model constructed the modern view for materials that compounds are produced by combination of atoms with the mass. This idea was supported by contemporary scientists. Gay-Lussac's law of combining volumes states that a simple integer ratio holds between the volumes of individual gases before reaction and the volumes of produced gases under the condition of constant pressure and temperature. This law means the volume of the gas is in proportion to the number of molecules under the above condition. Avogadro proposed the hypothesis that all gases of the same volume are the group of molecules with the same number. The mass of a material, A gram, consisting of an element with the atomic mass A is 1 g atom of the material. The mass of a compound, M gram, consisting of the molecular mass M is 1 g molecule or 1 mole. The number of atoms (molecules) included in the material of 1 g atom (molecule) is the common constant for all materials, called the Avogadro constant, N_A. The value of N_A is 6.022×10^{23} mol^{-1}.

The chemical element forms a series of characteristic lines peculiar to each element and is called element spectrum. The element absorbs the light of the same wavelength as the emitted light. Discharging a hydrogen atom contained in the gas discharge tube, a lot of spectral lines are observed using the spectrometer. A semiempirical formula giving the wavelength for the hydrogen spectrum was proposed by Balmer.

$$\frac{1}{\lambda} = R\left(\frac{1}{2^2} - \frac{1}{n^2}\right) \tag{3.1}$$

where $R = 1.09737 \times 10^7$ m^{-1} is called the Rydberg constant and n represents the integers greater than 2. This relationship was theoretically derived by Niels Bohr in 1913.

A few thousand volts is applied between two electrodes enclosed in a long and slender glass tube. Discharge occurs as the air pressure in the tube is reduced. At the pressure of 0.01 mmHg, a kind of radiation is ejected from the cathode. This is called the cathode ray. J.J. Thomson measured the ratio of its charge to the mass, e/m, called the specific charge. This is documented as the experimental discovery of the electron. The e/m value he obtained was about 1,700 times that of the hydrogen atom. The latest value is 1.7588×10^{11} C kg^{-1}.

3.2 Rutherford's Atomic Model

α-, β-, and γ-rays are used as probes to investigate the material structure. Rutherford and his students, Geiger and Marsden, worked on the α-particles' penetration of the material. When the collimated α-particle beam of 7.69 MeV hits the thin gold foil, penetrated α-particles are observed at various angles. Almost all of them deviate a little bit from the incident direction. However, large-angle scattering and back scattering rarely occurred. A very strong electric field or magnetic field is required to reverse the direction of high-speed particles. Rutherford thought that such a large-angle deflection is the evidence of the existence of a small and heavy nucleus having the positive charge. Light electrons in the atom move around the nucleus rapidly. An atom is almost an empty space. Therefore, almost all α-particles penetrated the foil without scattering. From these considerations, he calculated, assuming an α-particle is scattered, the Coulomb force from a point mass with positive charge. The Rutherford scattering formula was then derived. The calculated angular distributions were in good agreement with the experimental data. The nuclear radius for mass number A is approximately represented by

$$R \cong 1.3 A^{1/3} \times 10^{-15} \quad \text{m} \tag{3.2}$$

Therefore, the radius of a gold nucleus becomes 7.56×10^{-15} m. The radius of a gold atom is 1.79×10^{-10} m. The atomic model of Rutherford is called the solar system model.

3.3 Bohr's Quantum Theory

An object not moving with a constant speed and in the same direction is accelerated. From the viewpoint of classical mechanics, such an accelerated charge emits an electromagnetic wave. In this case, why is an atom of the Rutherford model stable? To explain this, Bohr proposed a theory that explains the spectrum of hydrogen. The theory of Bohr is based on three hypotheses:

1. The energy of an orbital electron is not continuous but one of discrete values peculiar to the atom, $E_1, E_2, E_3, ..., E_n$. These states, called the stationary states, do not emit light. The above energies are called the energy levels. The stationary state of the lowest energy is the ground state and the upper stationary states are the excited states.

2. An atom emits or absorbs the light when an electron jumps from a stationary state to another stationary state. If the electron state transfers from E_i to E_f, a photon of frequency v is emitted.

$$hv = E_i - E_f \qquad (3.3)$$

   This is called the Bohr's frequency condition.

3. An electron in the stationary state moves according to Newtonian mechanics. Bohr found that the correct energy level of an electron is obtained if the angular momentum of the electron around the nucleus is assumed to be the Planck's constant $\hbar$ times an integer, in which $\hbar = h/2\pi$. The circular motion of an electron around the hydrogen nucleus having the angular momentum of rmv, equal to $n\hbar$, is therefore

$$rmv = n\hbar \qquad (3.4)$$

   This is called the quantum rule. The integer n is called the quantum number.

The Coulomb force acting on an electron is equal to the centripetal force of the circular motion with the same speed; therefore,

$$m\frac{v^2}{r} = \frac{e^2}{4\pi\varepsilon_0 r^2} \qquad (3.5)$$

is obtained. From this equation, r becomes

$$r = \frac{4\pi\varepsilon_0 \hbar^2}{me^2} n^2 \qquad (3.6)$$

If $n = 1$, the radius is called Bohr's radius. This is usually noted a with the value of 0.529×10^{-10} m, and represents the radius of a hydrogen atom. The electron energy E is the sum of the kinetic energy and the Coulomb potential.

$$E = \frac{1}{2}mv^2 - \frac{e^2}{4\pi\varepsilon_0 r} \qquad (3.7)$$

Using Equations (3.5) and (3.6), the electron energy traveling the nth circular orbit becomes

$$E_n = -\frac{e^2}{8\pi\varepsilon_0 r} = -\frac{e^2}{8\pi\varepsilon_0 a}\frac{1}{n^2} = -\frac{13.6}{n^2} \text{ [eV]} \quad (n = 1, 2, 3, \ldots) \qquad (3.8)$$

This represents the energy level of a hydrogen atom. The light frequency ν accompanying the transition $n_i \rightarrow n_f$ is represented from Bohr's frequency condition,

$$h\nu = \frac{e^2}{8\pi\varepsilon_0 a}\left(\frac{1}{n_f^2} - \frac{1}{n_i^2}\right) \qquad (3.9)$$

Using the relationship of $c = \lambda\nu$,

$$\frac{1}{\lambda} = R\left(\frac{1}{n_f^2} - \frac{1}{n_i^2}\right) \qquad (3.10)$$

If $n_f = 2$, then the Balmer's series is derived.

$$R = \frac{e^2}{8\pi h\varepsilon_0 ac} = \frac{me^4}{8\varepsilon_0^2 h^3 c} \qquad (3.11)$$

is the theoretical equation of the Rydberg constant. In addition to the Balmer series, the Lyman series for $n_f = 1$, $n_i = 2, 3, 4, \ldots$, and the Paschen series for $n_f = 3$ were predicted by Bohr's theory. Bohr's quantum theory explained satisfactorily the spectrum of a hydrogen atom. The existence of the stationary state of an orbital electron in an atom was confirmed by the electron collision experiment carried out by Franck and Hertz. Bohr's theory mediated between the classical mechanics and the quantum mechanics. Nowadays this theory is called the old quantum theory.

EXAMPLE 3.1

Calculate the wavelengths of the first through third Lyman, Balmer, and Paschen lines.

SOLUTION 3.1

From Equations (3.10) and (3.11),

$$\frac{1}{\lambda} = R\left(\frac{1}{n_f^2} - \frac{1}{n_i^2}\right)$$

$$R = \frac{me^4}{8\varepsilon_0^2 h^3 c} = \frac{(9.109 \cdot 10^{-31}\ \text{kg})(1.602 \cdot 10^{-19}\ \text{c})^4}{8(8.854 \cdot 10^{-12}\ \text{cV}^{-1}\text{m}^{-1})^2(6.626 \cdot 10^{-34}\ \text{Js})^3(3 \cdot 10^8\ \text{ms}^{-1})}$$

$$= 1.096 \cdot 10^7\ \text{m}^{-1}$$

For the Lyman series, $n_f = 1$, $n_i = 2, 3, 4$; therefore,

$$\lambda_{n_i \to n_f} = \frac{1}{R\left(\dfrac{1}{n_f^2} - \dfrac{1}{n_i^2}\right)} = \frac{1}{R\left(1 - \dfrac{1}{n_i^2}\right)} = 121.64,\ 102.63,\ \text{and}\ 97.31\ \text{nm}$$

For the Balmer series, $n_f = 2$, $n_i = 3, 4, 5$:

$$\lambda_{n_i \to n_f} = \frac{1}{R\left(\dfrac{1}{4} - \dfrac{1}{n_i^2}\right)} = 656.84,\ 486.54,\ \text{and}\ 434.41\ \text{nm}$$

For the Paschen series, $n_f = 3$, $n_i = 4, 5, 6$:

$$\lambda_{n_i \to n_f} = \frac{1}{R\left(\dfrac{1}{9} - \dfrac{1}{n_i^2}\right)} = 1876.67,\ 1282.88,\ 1094.73\ \text{nm}$$

3.4 Quantum Mechanics

3.4.1 de Broglie Wave of Electrons

Light has the nature of wave-particle duality. DE Broglie considered that an electron regarded as a particle classically may have a wave nature. The wave-accompanying electron is called the electron wave. Generally, a wave accompanied by a particle is called a matter wave or de Broglie wave. When the energy E and the momentum p of the matter particle are given, the frequency and wavelength of the matter wave are written as

$$\nu = \frac{E}{h}, \quad \lambda = \frac{h}{p} \tag{3.12}$$

This is called the de Broglie's relation. Assuming the electron mass m, the charge e, and the accelerating voltage V, the electron velocity becomes $v = \sqrt{2eV/m}$. The de Broglie wavelength becomes

$$\lambda = \frac{h}{\sqrt{2meV}} \tag{3.13}$$

The de Broglie's relation was confirmed by the experiment of Davisson and Germer, in which electrons showed a diffraction phenomenon similar to that of x-rays.

EXAMPLE 3.2

What is the de Broglie wavelength of an electron accelerated by an electric potential of 54 V? Compare the electron's wavelength with the wavelength of a ball of 200 g moving with 50 m/s.

SOLUTION 3.2

For the electron:

$$\lambda = \frac{h}{\sqrt{2meV}} = \frac{(6.626 \cdot 10^{-34} \text{ Js})}{\sqrt{2 \cdot (9.109 \cdot 10^{-31} \text{ kg})(1.602 \cdot 10^{-19} \text{c}) \cdot V}}$$

$$= \frac{1.227 \text{ nm}}{\sqrt{V \text{(volt)}}} = \frac{1.227 \text{ nm}}{\sqrt{54}} = 1.67 \text{ nm}$$

For the ball:

$$\lambda = \frac{h}{mv} = \frac{(6.626 \cdot 10^{-34} \text{ Js})}{(0.2 \text{ kg})(50 \text{ m/s})} = 6.626 \cdot 10^{-26} \text{ nm}$$

The ball's wavelength is 26 orders of magnitude shorter than the electron's wavelength. (Therefore, the wave nature has little effect on ordinary objects.)

3.4.2 Uncertainty Principle

Light, which was considered a wave, exhibits a particle-like property, and an electron, which was considered a particle, exhibits a wave-like property. It is a fact that all matter exhibits both wave-like and particle-like properties; however, these are apparently not compatible. Consequently, the concept that light is a particle is restricted by its wave-like property. Similarly, the concept that a matter particle is a wave is restricted by its particle-like property. That means we cannot use the property of wave or particle for all matter without restriction. The rule that brings the restriction is called the uncertainty principle discovered by Heisenberg. The fundamental attributes of particles are the position in the space and the momentum, and that of waves is the wavelength. Momentum can be used in place of wavelength because of $p = h/\lambda$ between the wavelength λ and the momentum p. The uncertainty principle states both the position and the momentum of the matter particle cannot be determined at the same time. A relationship between the uncertainty of the position Δx and that for the momentum Δp holds:

$$\Delta x \cdot \Delta p \geq \hbar \qquad (3.14)$$

If Δp becomes smaller, Δx becomes larger. In opposition, if Δx becomes smaller, Δp must become larger. Accurate quantities for two parameters cannot be determined at the same time. This means wave-like and particle-like properties restrict each other, and unrestricted use of one property is not permitted. If we investigate the particle-like property for photons and electrons, their wave-like property is disturbed. There is a similar relationship between time Δt and energy ΔE, i.e.,

$$\Delta t \cdot \Delta E \geq \hbar \qquad (3.15)$$

EXAMPLE 3.3

What would be the annihilation time of an electron and a positron according to the uncertainty principle? How long would it take for a proton-antiproton pair?

SOLUTION 3.3

The positron and electron mass energy equivalent is 0.511 MeV and the Planck constant (h) is $4.136 \cdot 10^{-15}$ eV·s. Thus, from the uncertainty principle, the lifetime of an electron-positron pair is

$$\Delta t \geq \left(\frac{4.136 \cdot 10^{-15} \text{ eV} \cdot \text{s}}{2 \cdot 3.14159} \right) \cdot \frac{1}{2 \cdot 0.511 \cdot 10^6 \text{eV}} = 6.44 \cdot 10^{-22} \text{s}$$

The proton mass is 1,836 times larger than the electron mass; therefore, for a proton-antiproton pair it would take

$$\Delta t \geq \frac{6.44 \cdot 10^{-22} \text{ s}}{1836} = 3.5 \cdot 10^{-25} \text{ s}$$

3.4.3 Schrödinger Equation

In 1925, the theory of quantum mechanics was completed by Heisenberg and independently by Schrödinger. The formalism of Heisenberg is called matrix mechanics and that of Schrödinger wave mechanics. These two mechanics are equivalent in spite of different mathematical formalism. The same results are obtained. Here we give a simple introduction to wave mechanics. One of the objectives of quantum mechanics is to understand the concept of the wave-particle duality. It is assumed E is the electron energy, p the electron momentum, and v and λ the frequency and wavelength of the de Broglie wave, respectively. The relationships $v = E/h$ and $\lambda = h/p$ are transformed into a convenient form:

$$E = \hbar\omega, \quad p = \hbar k \tag{3.16}$$

where $\omega = 2\pi v$ and $k = 2\pi/\lambda$, which is called the wavenumber vector. Wave is a phenomenon that a wave quantity such as the displacement from the average level propagates in a direction without changing the spatial shape. If the quantity is assumed to be ϕ, the classical wave equation that describes the variation in time and space is represented by

$$\frac{1}{c^2}\frac{\partial^2 \phi}{\partial t^2} = \frac{\partial^2 \phi}{\partial x^2} + \frac{\partial^2 \phi}{\partial y^2} + \frac{\partial^2 \phi}{\partial z^2} \equiv \Delta\phi \tag{3.17}$$

in which Δ is the Laplace operator,

$$\Delta = \frac{\partial^2}{\partial x^2} + \frac{\partial^2}{\partial y^2} + \frac{\partial^2}{\partial z^2} \tag{3.18}$$

In the case of the sine wave propagating along the wavenumber vector k, ϕ is given by

$$\phi = A\sin(kr - \omega t) \tag{3.19}$$

where r is the position vector. For this ϕ,

$$\frac{\partial^2 \phi}{\partial t^2} = -\omega^2\phi \tag{3.20}$$

$$\Delta\phi = -k^2\phi \tag{3.21}$$

Inserting these two equations into Equation (3.17), $\omega = ck$ is obtained.

A complex solution can be considered in place of Equation (3.19), which is a real solution.

$$\phi = A\exp[i(kr - \omega t)] \tag{3.22}$$

In analogy with classical mechanics, it is assumed that the de Broglie wave is represented by an appropriate wave quantity ψ. This is called the wavefunction. The energy of an electron is $E = p^2/2m$ in the absence of a net force. Inserting this equation into Equation (3.16),

$$\omega = \hbar k^2/2m \tag{3.23}$$

is obtained. The function ψ represents the de Broglie wave and is written as in Equation (3.22). If so, $\Delta\psi = -k^2\psi$ holds. Therefore, Equation (3.23) is changed by multiplying ψ and using $E = \hbar\omega$,

$$-\frac{\hbar^2}{2m}\Delta\psi = E\psi \qquad (3.24)$$

This is called Schrödinger's wave equation, which is a fundamental equation to obtain the energy level or the energy eigenvalue. If there is a potential $U(x, y, z)$, the above equation becomes

$$-\frac{\hbar^2}{2m}\Delta\psi + U\psi = E\psi \qquad (3.25)$$

The solution of E derived from this equation for a hydrogen atom with the potential $U = -e^2/4\pi\varepsilon_0 r$ agrees exactly with the solution of Bohr's theory (Equation 3.8).

3.4.4 Wavefunction

In Bohr's theory an electron exists just on a circular orbit with a certain radius, while the solution of the Schrödinger equation extends to around this circular orbit. If an electron exists somewhere in the space, its mass and charge exist there. Namely, an electron is an individual particle. The distributed wavefunction in space and the individuality of an electron is inconsistent. Bohr proposed an interpretation for this question. The probability $p(x, y, z, t)$ that a particle exists in a small volume $dxdydz$ at a point (x, y, z) at a time t is given by

$$p(x,y,z,t)\,dxdydz = |\psi(x,y,z,t)|^2\,dxdydz \qquad (3.26)$$

ψ does not depend on t for the stationary state; therefore, the probability over the whole space becomes

$$\iiint |\psi(x,y,z)|^2\,dxdydz = 1 \qquad (3.27)$$

It is understood that the wavefunction decides the statistical behavior of a particle.

EXAMPLE 3.4

What is the general solution of the wavefunction for a constant potential U? Consider a one-dimensional problem.

SOLUTION 3.4

From Equation (3.25), the Schrodinger equation for a one-dimensional problem is given by

$$-\frac{\hbar^2}{2m}\frac{\partial^2 \Psi}{\partial x^2} + U\Psi = E\Psi$$

$$\frac{\partial^2 \Psi}{\partial x^2} = -\frac{2m}{\hbar^2}(E-U)\Psi$$

where A and B are constants, and $k = \sqrt{\frac{2m}{\hbar^2}(E-U)}$.
   Therefore:

If $k > 0$, $\Psi = A\exp(-ikx) + B$ (wave form)

If $k \leq 0$, $\Psi = A\exp(\pm kx) + B$

To have a finite solution at infinitely large x, the solution for this case is $\Psi = A\exp(-kx) + B$ (the wavefunction is decreasing with the distance $x \Rightarrow$ tunneling).

3.5 Atomic Structure

3.5.1 Electron Orbit

Quantum mechanics provides the strict solution just for the hydrogen atom ($Z = 1$). However, it is impossible to provide strict solutions for atoms with $Z \geq 2$. Adequate approximations are applied to these atoms. The outline for atomic structure obtained by these methods is described here. The energy level of an atom is specified by a set of three quantum numbers (n, l, m). The principal quantum number n is allowed $n = 1, 2, 3, \ldots$, corresponding to the magnitude of the electron orbit. The azimuthal quantum number l, which represents the magnitude of angular momentum of an electron, is allowed, $l = 0, 1, \ldots, n - 1$, for a given n. The alternative notation is $s, p, d, f, \ldots$. The third quantum number m is called the magnetic quantum number, which was introduced to explain the split of spectrum under the magnetic field (Zeeman effect). The values of m corresponding to the components of angular momentum are allowed $m = 0, \ldots, \pm(l-1), \pm l$. A state specified ($n$, l, m) is called the orbit. According to quantum mechanics, the distribution of electrons extends like a cloud with a spherical shell. The spherical shells are called K-shell, L-shell, M-shell, N-shell, etc., corresponding to $n = 1, 2, 3, 4, \ldots$. Energy levels depend on n and l but not m. The value of energy level, E_{nl}, is higher as n is larger for the same l and is higher as l is larger for the same n. On the E_{2p} state, three states of $m = -1$,

0, 1 are allowed even if the same energy. Such a state is called a degenerate energy level.

3.5.2 Pauli's Exclusion Principle

In consideration of atomic structure Pauli's exclusion principle plays an important role. The number of electrons included in an orbit specified by (n, l, m) is restricted to at most two. Moreover, if two electrons are included in an orbit, the direction of electron spin should be reversed. According to this principle, the ground state is realized by filling electrons starting with the lowest level. Table 3.1 shows the electron configuration for the ground state of atoms up to Kr ($Z = 36$). For example, the electron configuration is represented as $(1s)^2$ for He ($Z = 2$). The Li ($Z = 3$) has such a configuration of one electron at the 2s orbit added to the closed shell of He. Based on the energy level diagram, the periodic table of elements and their chemical properties are understood.

TABLE 3.1

Electron Configuration for the Ground State of Atoms up to Kr ($Z = 36$)

| | | Energy Level | | | | | | | |
| | | K | L | | M | | | N | |
Z	Element	1s	2s	2p	3s	3p	3d	4s	4p
1	H	1							
2	He	2							
3	Li	2	1						
4	Be	2	2						
5	B	2	2	1					
6	C	2	2	2					
7	N	2	2	3					
8	O	2	2	4					
9	F	2	2	5					
10	Ne	2	2	6					
11	Na	2	2	6	1				
12	Mg	2	2	6	2				
13	Al	2	2	6	2	1			
14	Si	2	2	6	2	2			
15	P	2	2	6	2	3			
16	S	2	2	6	2	4			
17	Cl	2	2	6	2	5			
18	Ar	2	2	6	2	6			

(continued)

TABLE 3.1 (CONTINUED)

Electron Configuration for the Ground State of Atoms up to Kr (Z = 36)

		K	L		M			N	
Z	Element	1s	2s	2p	3s	3p	3d	4s	4p
19	K	2	2	6	2	6		1	
20	Ca	2	2	6	2	6		2	
21	Sc	2	2	6	2	6	1	2	
22	Ti	2	2	6	2	6	2	2	
23	V	2	2	6	2	6	3	2	
24	Cr	2	2	6	2	6	4	1	
25	Mn	2	2	6	2	6	5	2	
26	Fe	2	2	6	2	6	6	2	
27	Co	2	2	6	2	6	7	2	
28	Ni	2	2	6	2	6	8	2	
29	Cu	2	2	6	2	6	10	1	
30	Zn	2	2	6	2	6	10	2	
31	Ga	2	2	6	2	6	10	2	1
32	Ge	2	2	6	2	6	10	2	2
33	As	2	2	6	2	6	10	2	3
34	Se	2	2	6	2	6	10	2	4
35	Br	2	2	6	2	6	10	2	5
36	Kr	2	2	6	2	6	10	2	6

EXAMPLE 3.5

What is the electron configuration of an oxygen atom? How many L-electrons does it have?

SOLUTION 3.5

An oxygen atom has in total eight electrons: $(1s)^2(2s)^2(2p)^4$, of which six electrons are the L-electrons ($n = 2$).

3.6 Summary

1. Dalton's law in chemical reaction (the law of definite proportions and the law of multiple proportions) is a pioneer of modern atomic theory.
2. The spectrum of a hydrogen atom was explained by Bohr's quantum theory.

3. In quantum mechanics, the square of the absolute value of a wave-function represents the existence probability of particles.

4. The uncertainty principle states that the position and the momentum of a particle cannot be determined accurately at the same time.

5. The energy level of an atom is specified by the principal quantum number n, the azimuthal quantum number l, and the magnetic quantum number m. One state specified (n, l, m) is called the orbit.

QUESTIONS

1. Describe ways in which atoms of a gas can be excited to emit light with frequencies that are of the characteristic line spectrum of the element involved.
2. Under what condition does an electron remain in an orbit without emitting energy?
3. Compare binding energy and ionization potential of an α-particle.
4. What is the value of the ionization potential of hydrogen?
5. Using the uncertainty principle, explain numerically why the hydrogen atom is stable rather than an electron merging into a proton.
6. Using the uncertainty principle, explain why a hydrogen atom exists as H_2 rather than H.
7. Calculate the radius of a Bohr atom.
8. Calculate the velocity of an electron in the first Bohr orbit.
9. What is the ratio of the Bohr velocity to the velocity of light known as? Obtain its numerical value of α.
10. Equation $\Delta t \cdot \Delta E \approx h/2\pi$ permits a temporary violation of conservation of energy. How long is a violation of 0.03 eV tolerated?
11. What is the closest approach an α-particle of 5.3 MeV energy can make to a water molecule?
12. Calculate the number and show the configuration of the number of electrons allowed in the M-shell.

For Further Reading

Johns HE, Cunningham JR. 1974. *The Physics of Radiology*, 3rd ed. Springfield, IL: Charles C. Thomas Publisher.

Turner JE. 1995. *Atoms, Radiation, and Radiation Protection*, 2nd ed. New York: John Wiley & Sons.

4

Atomic Nucleus

4.1 Constituents of Nucleus

A nucleus of the atomic number Z and the mass number A consists of Z protons and $N = A - Z$ neutrons. The mass number A is the sum of nucleons (protons and neutrons) in a nucleus. A nucleus identified by A and Z is called a nuclide. The notation of a nuclide is $^A_Z E$, in which E is an element, for example, $^{16}_8 O$. At the present time, there exist about 300 stable nuclides and about 1,700 unstable radioactive nuclides. The classification of nuclides is as follows:

1. Isotope has the same Z. Chemical properties are the same because the number of orbital electrons and that of electron configurations are the same. Radioisotopes are unstable in energy; therefore, a nuclide decays to another nuclide by emitting radiation. This ability is called radioactivity.

2. Isobar has the same A.

3. Isotone has the same N.

4. Isomer has the same Z and the same N.

4.2 Binding Energy of Nucleus

There is a relationship between atomic mass unit (u) and energy (MeV). From the definition the ^{12}C atom is 12 u. Because its gram-atomic weight is 12 g,

$$1u = 1/(6.022 \times 10^{23}) = 1.6605 \times 10^{-27} \quad kg \tag{4.1}$$

This corresponds to 931.5 MeV using Einstein's formula, $E = mc^2$. If the mass ΔM is lost, the released energy Q becomes $(\Delta M)c^2$. The mass of a neutral atom specified by A and Z is $M(A, Z)$, and the mass of a hydrogen atom, M_H. If the electron binding energy is ignored, the total binding energy of a nuclide (A, Z), B, is given by

$$B = \{ZM_H + (A - Z)M_n - M(A, Z)\}c^2 = \Delta Mc^2 \qquad (4.2)$$

where ΔM is called the mass defect. Using the atomic mass units, M_H and M_n are 1.007825 u and 1.008665 u, respectively. An alternative representation, the mass deviation can be used to obtain B.

$$\Delta = M - A \qquad (4.3)$$

in which M is the mass and A is an integer mass number. That for ^{12}C becomes $\Delta = 0$ from the definition. The binding energy means the decrease of the total energy when a nucleus is formed by combining isolated free nucleons. In other words, that is the energy released when nucleons bind as a nucleus. The value of B for ^{4}He atom becomes

$$B = [(2 \times 1.007825 + 2 \times 1.008665) - 4.002603] \times 931.5 = 28.30 \quad \text{MeV}$$

$$= 7.07 \text{ MeV/n} \qquad (4.4)$$

The energy of chemical reaction is at most a few eV. For ^{4}He, an energy of about 10 million times the chemical reaction is dissipated for binding the nucleons. The binding energy of a nucleon is denoted by B/A. Figure 4.1 shows B/A for the stable nuclides as a function of A. In the region of $A \geq 40$,

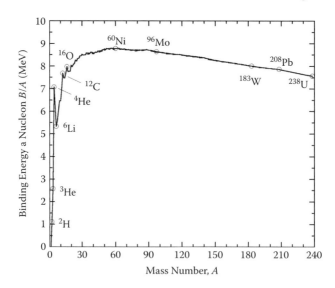

FIGURE 4.1
Mean binding energy of a nucleon.

for example, for ^{56}Fe $= 8.8$, B/A is almost constant, ~8 MeV/n. This is called the saturation of the binding energy. The binding energy is the largest, namely, the most stable, for nuclides of $A = 55 \sim 60$. If an unstable nucleus is converted to a more stable nucleus, enormous energy is released. One method is the nuclear fusion by making a helium nucleus from hydrogen isotopes. Another example is the nuclear fission, by breaking a heavy nucleus like uranium into two light nuclei.

EXAMPLE 4.1

Obtain the absolute value of the atomic mass unit for 1 mole of ^{12}C (0.012 kg).

SOLUTION 4.1

0.012 kg $= N_A.12\, m_u$; therefore, $1u = 1.661 \times 10^{-17}$ kg $\equiv 931.48$ MeV.

4.3 Nuclear Models

4.3.1 Liquid Drop Model

To understand various overall properties of nuclei, various models have been created, such as the shell model and the liquid drop model. The liquid drop model can satisfactorily explain many of the nuclear phenomena. This model arose from saturation phenomena of the nuclear density and the binding energy being analogous with the liquid drop. Here, the nuclear binding energy corresponds to the intermolecular force in the liquid. Nuclear reactions and nuclear fission are explained by the liquid drop model. When a projectile enters the nucleus, the energy enhances the temperature of the nucleus and particles are emitted similar to evaporation. Nuclear fission looks like separation into two liquid drops when energy is given to a large liquid drop. Weizsäcker and Bethe proposed a semiempirical mass formula representing the mass $M(A, Z)$ for a neutral atom of $A > 15$.

$$M(A,Z) = ZM_H + (A - Z)M_n - a_V A + a_S A^{2/3} + a_C \frac{Z^2}{A^{1/3}} + a_a \frac{(A/2 - Z)^2}{A} + \delta(A,Z)$$

$$(4.5)$$

where the parameters a_V, a_S, a_C a_a, and $\delta(A,Z)$ are determined to fit to the measured data.

EXAMPLE 4.2

Use the Bethe-Weizsäcker mass formula to calculate the binding energy of a U-235 nucleus, given that

$$a_V = 15.75 \text{ MeV}, a_S = -17.8 \text{ MeV}, a_C = 0.711 \text{ MeV}, a_a = 23.7 \text{ MeV, and}$$

$$\delta = \begin{cases} -\dfrac{11.18 \text{ MeV}}{\sqrt{A}}, & Z \text{ and } N \text{ are even numbers} \\ 0, & \text{either } Z \text{ or } N \text{ are even numbers} \\ +\dfrac{11.18 \text{ MeV}}{\sqrt{A}}, & Z \text{ and } N \text{ are odd numbers} \end{cases}$$

Calculate B/A and the nuclear mass.

SOLUTION 4.2

$$B = [Z \cdot M_H + (A - Z)M_n] - M(A, Z)$$

$$= (15.75 \text{ MeV}) \cdot 235 - (17.8 \text{ MeV}) \cdot 235^{2/3} - (0.711 \text{ MeV}) \cdot \frac{92^2}{235^{1/3}}$$

$$- (23.7 \text{ MeV}) \cdot \frac{(235 - 2 \cdot 92)^2}{235} + 0$$

$$= 1785.9 \text{ MeV}$$

$$B/A = 1785.9 \text{ MeV}/235 = 7.6 \text{ MeV/n}$$

$$M = [Z \cdot M_H + (A - Z)M_n] - B$$

$$= [92 \cdot (1.007825 \text{ u}) + (235 - 92) \cdot (1.008665 \text{ u})] - \frac{1785.9 \text{ MeV}}{931.5 \text{ MeV/u}}$$

$$= 235.0418 \text{ u}$$

4.3.2 Shell Model

The shell model originates from the accumulation of a lot of experimental data showing the existence of a closed shell in a nuclear structure. In the case that the number of protons Z or that of neutrons N is 2, 8, 20, 28, 50, 82, and 126, the nucleus lies especially stable. Therefore, it is thought that the nucleus forms the closed shell for Z or N of these numbers. These numbers are called the magic number. The liquid drop model could not explain the magic number. The shell model was proposed to settle this problem. Discrete magic numbers remind us that atoms, like inert gases, are stable because of the closed shell structure of the orbital electrons. For atoms, the electron configuration of the ground state was obtained by filling electrons in order, starting with the lowest energy level. A similar method was applied to the nucleus. This model assumes individual nucleons move independently in the average potential replacing the nuclear force. In addition, Mayer and Jensen (1955) introduced the spin-orbit coupling potential. They succeeded in explaining all magic numbers as the closed shell. Their potential is given by

$$V(r) = V_0(r) + f(r)\boldsymbol{L} \bullet \boldsymbol{S} \qquad (4.6)$$

in which $\boldsymbol{L}$ and $\boldsymbol{S}$ are the orbital angular momentum and spin quantum number, respectively. The potential $V_0(r)$ was assumed to be an intermediate type between the harmonic oscillator and the well. The spin-orbit coupling potential generates the splitting of the energy levels. Figure 4.2 shows the energy levels calculated by the shell model. Because the value of $f(r)$ is negative, the level with the total angular momentum $I = L + 1/2$ is lower than the level with $I = L - 1/2$. Each level is represented as nL_I, where n is the principal quantum number. The occupancy number in a level I is $2I + 1$ for the same kind of particle. The total number becomes 2, 8, 20, ..., at the levels $1s_{1/2}$, $1p_{1/2}$, $1d_{3/2}$, ..., when nucleons are filled in order from the lowest level. It is found that the energy gap between these levels and their upper levels is large in comparison with other gaps; therefore, these levels form the closed shell.

4.3.3 Collective Model

The electric quadrupole moment of the nucleus is determined by distributions of charge and current in the nucleus. We understand the nuclear size, the shape, and the density from this. The shell model could not explain this quantity. Bohr and Mottelson proposed the collective model in which the shape of a closed shell is not spherical symmetry but deformed spheroid. The deformed nucleus rotates around the axis perpendicular to the symmetry axis of a spheroid if energy is given. Consequently, rotational levels of the quantized rigid top may appear. The electric quadrupole moment and the excited levels called the rotational band calculated using the theory of Bohr-Mottelson were in agreement with the experimental values. In addition, the collective model explained the vibrational excited mode assuming the vibration of nuclear matter.

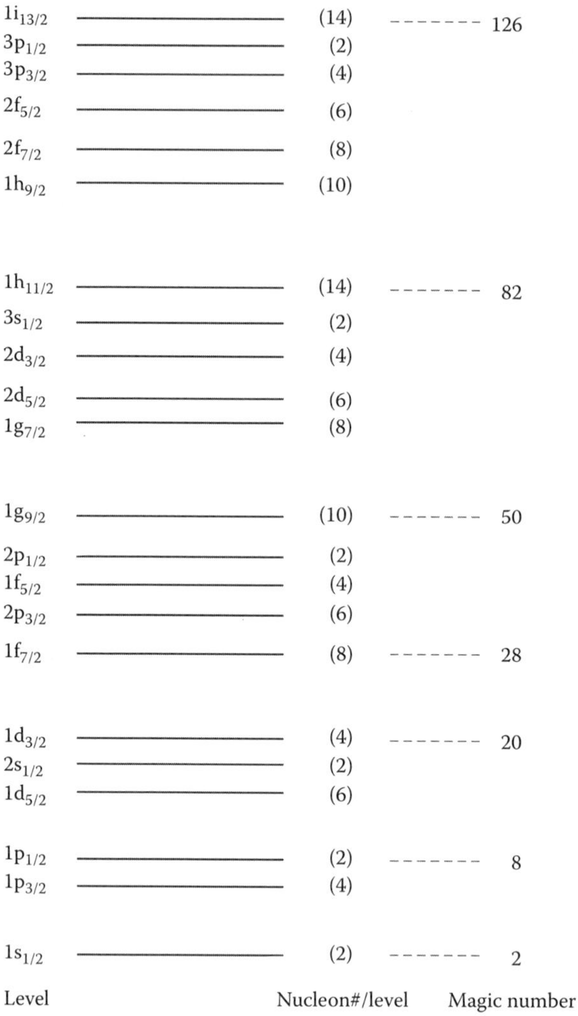

Level		Nucleon#/level		Magic number
$1i_{13/2}$	————————	(14)	— — — — — —	126
$3p_{1/2}$	————————	(2)		
$3p_{3/2}$	————————	(4)		
$2f_{5/2}$	————————	(6)		
$2f_{7/2}$	————————	(8)		
$1h_{9/2}$	————————	(10)		
$1h_{11/2}$	————————	(14)	— — — — — —	82
$3s_{1/2}$	————————	(2)		
$2d_{3/2}$	————————	(4)		
$2d_{5/2}$	————————	(6)		
$1g_{7/2}$	————————	(8)		
$1g_{9/2}$	————————	(10)	— — — — — —	50
$2p_{1/2}$	————————	(2)		
$1f_{5/2}$	————————	(4)		
$2p_{3/2}$	————————	(6)		
$1f_{7/2}$	————————	(8)	— — — — — —	28
$1d_{3/2}$	————————	(4)	— — — — — —	20
$2s_{1/2}$	————————	(2)		
$1d_{5/2}$	————————	(6)		
$1p_{1/2}$	————————	(2)	— — — — — —	8
$1p_{3/2}$	————————	(4)		
$1s_{1/2}$	————————	(2)	— — — — — —	2

FIGURE 4.2
Energy level scheme of a nuclear shell model.

4.4 Nuclear Reaction

4.4.1 Characteristics

When particles collide with the nucleus, various phenomena arise by the interaction between the target nucleus and the projectile. Those are generally called a nuclear reaction. For the purposes of study of a nucleus and nuclear transformation, nuclear reactions using the projectiles of protons, neutrons, heavy ions, and electrons and photons, too, are performed. After the interaction of the projectile a with the target nucleus A, the emitted particle b and the residual nucleus B are produced. This reaction is written as $a + A \rightarrow B + b$ or $A(a, b)B$. There are various kinds of interactions. In a reaction the target nucleus is converted into another nucleus. Inelastic scattering accompanies the change of inner state of the target nucleus. Elastic scattering changes the direction of the projectile. Classification of a direct reaction or compound reaction is made from the difference in the reaction mechanism.

4.4.2 Cross Section

Projectiles incident on the target do not always induce reaction. Even if reaction occurs, not a specific reaction but many different reactions occur. A specific reaction occurs with a certain probability. This probability is called the cross section. It is assumed that a thin target with the volume V m³, including a nucleus with a density of n m⁻³, is put in the uniform stream of the incident particles. If the fluence rate of the particle is ϕ m⁻²s⁻¹, the number of reaction in the unit time, N s⁻¹, is given by

$$N = \sigma \phi n V \qquad (4.7)$$

in which σ represents the cross section. Its unit is m² or barn (b) (1b = 10^{-28} m²). A partial cross section is for a specific reaction type, and a total cross section is the sum of all possible reactions. The value of a cross section does not mean a simple geometrical area but depends on the incident energy and the nuclear structure of the target. The rough value for the nuclear reaction is estimated at $\sim 10^{-30}$ m² because of the diameter of nucleus $\sim 10^{-15}$ m. The differential cross section $d\sigma/d\Omega$ is defined to describe the angular distribution of the ejected particles. The differential cross section in energy $d\sigma/d\varepsilon$ describes the energy distribution.

EXAMPLE 4.3

5 MeV neutrons with the fluence rate 10 m⁻² s⁻¹ are emitted to a water tank. Calculate number of neutron interactions in 1 m³ water.

SOLUTION 4.3

$$\sigma_0 = 1.55 * 10^{-28}\,\text{m}^2,$$

$$\sigma_H = 1.5 * 10^{-28}\,\text{m}^2$$

$$N = \sigma_0 \phi \frac{\rho}{A} n_A f_o + \sigma_H \phi \frac{\rho}{A} n_A f_H$$

$$= (1.55 * 10^{-28}\,\text{m}^2) \cdot (10\text{m}^{-2}\text{s}^{-1}) \cdot \left(\frac{1000}{0.016} \cdot 6.02 * 10^{23} \right) \cdot \frac{1}{3}$$

$$+ (1.5 * 10^{-28}\,\text{m}^2) \cdot (10\text{m}^{-2}\text{s}^{-1}) \cdot \left(\frac{1000}{0.001} \cdot 6.02 * 10^{23} \right) \cdot \frac{2}{3}$$

$$= 621(\text{s}^{-1})$$

4.4.3 Threshold Value of Reaction

We consider a reaction $X + a \rightarrow Y + b$ or $X(a, b)Y$. The energy released in the reaction is called the Q value of reaction. If $Q < 0$, namely, energy is absorbed, the reaction is called the endothermic reaction. If $Q > 0$, it is called the exothermic reaction. The reaction including Q is written as

$$X + a \rightarrow Y + b + Q \qquad (4.8)$$

Extra energy is required in order to cause the endothermic reaction. The threshold energy of reaction, E_{th}, is calculated from the kinematical relationships. Assuming the head-on collision of the projectile a, with a mass M_a on the target M_X in rest, M_Y and M_b are produced after collision. The difference of the rest energy becomes

$$Q = M_a + M_X - \left(M_b + M_Y\right) \qquad (4.9)$$

If an endothermic reaction, the sign of Q is minus. Since the total energy is conserved,

$$E_a = E_Y + E_b - Q \qquad (4.10)$$

is obtained in which E represents the kinetic energy. From the momentum conservation,

$$p_a = p_Y + p_b \qquad (4.11)$$

is obtained. If E_Y and E_b are deleted from the above equations, the threshold energy E_a is obtained

$$E_{th} = -Q\left(1 + \frac{M_a}{M_Y + M_b - M_a}\right) \qquad (4.12)$$

For positively charged projectiles, the actual threshold is greater than E_{th} due to the Coulomb's repulsive force. The projectile needs acceleration to some extent even if $Q > 0$. Various reactions are easily caused for the case of neutrons without the Coulomb barrier.

EXAMPLE 4.4

Calculate the Q value and the threshold energy of the projectile for the reactions $^{13}C(d, t)^{12}C$ and $^{14}C(p, n)^{14}N$. Are these reactions endothermic or exothermic?

Atomic mass can be found at http://t2.lanl.gov/data/astro/molnix96/massd.html.

SOLUTION 4.4

For $^{13}C(d, t)^{12}C$:

$$Q = (13.00335503 + 2.01410174)\ u - (3.01604939 + 12)\ u$$

$$= 0.001407\ u = 1.3157\ MeV$$

Exothermic reaction: No threshold energy is required.

For $^{14}C(p, n)^{14}N$:

$$Q = (14.00324154 + 1.007825)\ u - (1.008665 + 14.00307369)\ u$$

$$= -0.00063\ u = -0.59\ MeV$$

Endothermic reaction: The threshold energy is

$$E_{th} = 0.59\ MeV \cdot \left(1 + \frac{1.007825}{14.00307369 + 1.008665 - 1.007825} \right)$$

$$= 0.63\ MeV$$

4.5 Nuclear Fission

When slow neutrons hit $^{235}_{92}U$, two more light nuclides are produced. This phenomenon is called nuclear fission. The binding energy B/A is ~7.6 MeV in the uranium region ($A \sim 240$). On the other hand, that value is ~8.5 MeV in the region of $A \sim 120$. It is estimated, therefore, the released energy is $240 \times (8.5 - 7.6) = 210$ MeV. A major part of this is converted into the kinetic energy of the fission products. The fission products do not separate in two right away but show an unsymmetrical pattern in the mass distribution. The fission fragments usually have excess neutrons; therefore, successive β-decays occur. The number of emitted neutrons in a fission reaction was confirmed. That was important from the viewpoint of the chain reaction of nuclear fission. Table 4.1 lists the average distribution of energies ~200 MeV released by the fission of ^{235}U. Two kinds of neutrons are ejected from the fission fragments.

TABLE 4.1

Average Distribution of Energies Released by Nuclear Fission of ^{235}U

Kinetic energy of fission fragments ($A \sim 96$, $A \sim 140$)	165 ± 5 MeV
Kinetic energy of fission neutrons ($2 \sim 3$)	5 ± 0.5
Energy of prompt γ-rays (~5)	6 ± 1
Energy of β-rays from fission fragments (~7)	8 ± 1.5
Energy of γ-rays from fission fragments (~7)	6 ± 1
Energy of neutrinos from fission fragments	12 ± 2.5
Total energy of nuclear fission	202 ± 6 MeV

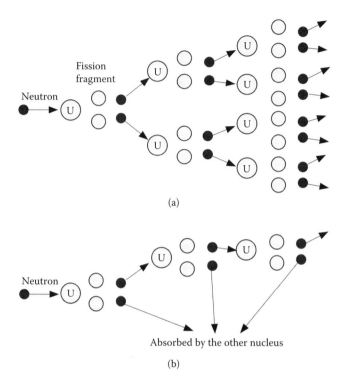

FIGURE 4.3
Schemes of chain reaction: (a) uncontrolled and (b) controlled.

One is the prompt neutrons ejected immediately ($\sim 10^{-12}$ s), and the other is the delayed neutrons after 1 s ~ 1 min. When the fission of ^{235}U occurs by a slow neutron, the mean number of emitted neutrons is 2.47. This value suggests successive chain reaction due to produced neutrons. Figure 4.3 shows the schemes of chain reactions, uncontrolled and controlled. If all emitted two more neutrons are absorbed by ^{235}U, the chain reaction abruptly increases and results in an explosion (the atomic bomb). The controlled chain reaction always uses one neutron under the constant rate in time.

4.6 Nuclear Fusion

The formation of the intermediate nuclei combining the light nuclei is called nuclear fusion, in which nuclear energy is also released. The most efficient nuclear fusion is

$$^2D + {}^2D \rightarrow {}^3He + n + 3.27 \text{ MeV},$$

$$^2D + {}^2D \rightarrow {}^3T + p + 4.03 \text{ MeV}$$

$$(4.13)$$

Moreover, such a fusion occurs using ^{3}T,

$$^2D + {}^3T \rightarrow {}^4He + n + 17.58 \text{ MeV} \tag{4.14}$$

EXAMPLE 4.5

Calculate the Q value for the $^2D + {}^3T \rightarrow {}^4He + n$ fusion reaction.

SOLUTION 4.5

$$Q = (3.016049 + 2.014102 - 4.002604 - 1.008665) * 931.5 = 17.58 \text{ MeV}$$

It is important for us to control these nuclear fusions. At very high temperature, thermonuclear reactions are sustained because nuclei in matter having thermal energy collide with each other. By the way, the average energy of a thermal motion is kT. The temperature T becomes 1×10^7 K for 1 keV. The problem is to keep such a high temperature as $10^7 \sim 10^9$ K. This problem has not yet been solved.

The origin of radiation energy of the sun is the following nuclear fusion:

$$^1H + {}^1H \rightarrow {}^2H + e^+ + \nu$$

$$e^+ + e^- \rightarrow 2\gamma$$

$$^2H + {}^1H \rightarrow {}^3He + \gamma \tag{4.15}$$

$$^3He + {}^3He \rightarrow {}^4He + 2{}^1H$$

From the above reactions,

$$4{}^1H + 2e^- \rightarrow {}^4He + 2\nu + 6\gamma \tag{4.16}$$

is obtained. Consequently ^{4}He is formed from four protons accompanied by γ-ray emission of 27 MeV.

4.7 Summary

1. A nucleus of the atomic number Z and the mass number A consists of Z protons and $A - Z$ neutrons.

2. Radioactive isotopes convert into a different element emitting radiations. Radioactivity means this capability.

3. The binding energy represents a reduction of the whole energy when a nucleus is formed by binding of isolated free nucleons.

4. The shell model of a nucleus explains the magic numbers 2, 8, 20, 28, 50, …, for Z or N at which the nucleus is stable.

5. The energy released in the nuclear reaction is called the Q value. The endothermic and exothermic reactions occur for $Q < 0$ and $Q > 0$, respectively.

QUESTIONS

1. Complete the table.

Radiation	nature	Atomic mass (u)	Charge	Energy (MeV)	Velocity (c)	Range in Air	Range in Water	Range in Al	Ionization relative to α-particle
λ									
β									
α									

2. The Weizsäcker semiempirical mass formula is given by Equation 4.5
   a. Identify each term in the equation.
   b. Derive an equation in MeV for M according to even and odd A.
   c. What is the value of δ?
3. What is the intrinsic parity of (a) photon, (b) electron, (c) proton, and (d) neutron? In the case of a photon, comment on the origin of the photon.
4. What is the work done in MeV for two protons separated by 5 F (1 fermi = 10^{-15} m). How does this energy compare with the nuclear force between the two protons?

For Further Reading

Mayer M, Jensen J. 1955. *Elementary Theory of Nuclear Shell Structure*. New York: John Wiley & Sons.

Turner JE. 1995. *Atoms, Radiation, and Radiation Protection*, 2nd ed. New York: John Wiley & Sons.

5

Radioactivity

5.1 Types of Radioactivity

5.1.1 α-Decay

An unstable atomic nucleus spontaneously loses energy by emitting ionizing particles and radiation. This process is called the decay of the nucleus.
Almost all heavy elements of $Z \geq 83$ spontaneously emit α-rays, which is
the nucleus of ^{4}He. Both the atomic number Z and the number of neutron N
decrease by 2. An example of the α-decay is

$$^{238}_{92}U \rightarrow \ ^{234}_{90}Th + \alpha \tag{5.1}$$

The energy released in this decay, namely, the Q value, is given by

$$Q = M(A,Z) - [M(A-4,Z-2) + M(2,2)] \tag{5.2}$$

where M represents the atomic mass. This energy is shared by the α-particle
and the recoil nucleus. From the kinematical relationship between them,
it is understood that the kinetic energy of the α-particle is discrete. Even
α-particles with lower energy than the barrier of Coulomb potential can
jump out of nucleus owing to the tunnel effect. Figure 5.1 shows the potential
for α-particles and the scheme of the tunnel effect. The Coulomb potential is

$$U(r) = \frac{2Ze^2}{4\pi\varepsilon_0 r} \tag{5.3}$$

If $r = 1 \times 10^{-12}$ cm, $Z = 85$, then the magnitude of the barrier $U \approx 25$ MeV
is obtained. This value is much higher than ordinary α-particle energies of
a few MeV. α-Particles cannot pass through the barrier from the classical
viewpoint. The quantum theory for the tunnel effect settled the problem.
The penetration rate P is given by

$$P = e^{-G}, \quad G = \frac{2}{\hbar} \int_R^b \sqrt{2m_\alpha(U(r)-E)} \ dr \tag{5.4}$$

in which R and h are the radii, as shown in the figure. The value of P is nonzero.

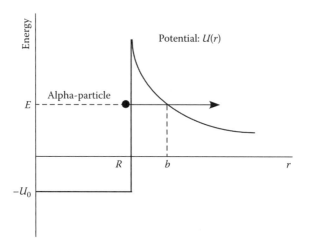

FIGURE 5.1
Potential energy of a nucleus for an α-particle.

For example, α-decay of $^{230}_{90}$Th is shown as

$$^{230}_{90}\text{Th} \rightarrow {}^{226}_{88}\text{Ra} + {}^{4}_{2}\text{He} \tag{5.5}$$

Figure 5.2 shows the decay scheme of a decay of ^{230}Th. The released energy in this decay, $Q = 4.771$ MeV, is the nuclear mass difference between Th and Ra + He.

$$Q = M_{\text{Th}} - \left(M_{\text{Ra}} + M_{\alpha}\right) \tag{5.6}$$

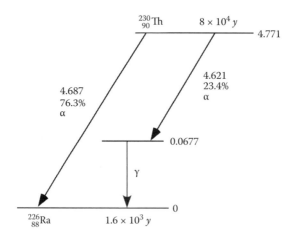

FIGURE 5.2
Decay scheme of ^{230}Th (Lederer et al. 1978).

This energy is shared between the α-particle and the recoiled radium. The magnitude of momentum of Ra is equal to that of α, and they go to opposite directions. Assuming the velocities V and v for Ra and α, respectively, the momentum conservation is represented by

$$M_\alpha v = M_{Ra} V \tag{5.7}$$

The relationship for energy is given by

$$Q = \frac{1}{2} M_\alpha v^2 + \frac{1}{2} M_{Ra} V^2 \tag{5.8}$$

From these equations, the energies of α-particle and radium become

$$E_\alpha = \frac{M_{Ra} Q}{M_\alpha + M_{Ra}}, \quad E_{Ra} = \frac{M_\alpha Q}{M_\alpha + M_{Ra}} \tag{5.9}$$

Inserting $Q = 4.771$ MeV into the first equation, $E_\alpha = 4.688$ MeV is obtained. It is understood that the α-particle has the discrete energy.

EXAMPLE 5.1

Calculate the value and energy of the α-particle and the daughter nucleus, originating from the decay of $^{226}_{88}$Ra.

SOLUTION 5.1

The decay of $^{226}_{88}$Ra is

$$^{226}_{88}\text{Ra} \rightarrow {}^{222}_{86}\text{Rn} + {}^{4}_{2}\text{He}$$

$$Q = M\left({}^{226}_{88}\text{Ra}\right) - M\left({}^{222}_{86}\text{Rn}\right) - M\left({}^{4}_{2}\text{He}\right)$$

$$= (226.0254 - 222.0176 - 4.0026) \text{ u}$$

$$= 0.0052 \text{ u} = 0.0052 \cdot 931.5 \text{ MeV} = 4.8438 \text{ MeV}$$

The α-particle has the energy of

$$E_\alpha = \frac{M_{Rn} Q}{M_{Rn} + M_\alpha} = \frac{222.0176 \cdot 4.8438 \text{ MeV}}{222.0176 + 4.0026} = 4.758 \text{ MeV}$$

$$E_{Rn} = (4.8438 - 4.758) \text{ MeV} = 0.0858 \text{ MeV}$$

5.1.2 β⁻ Decay

β⁻ decay is a phenomenon that a neutron changes into a proton following the emission of an electron (β⁻) and an antineutrino.

$$n \rightarrow p + e^- + \bar{\nu}_e \tag{5.10}$$

When this decay occurs within a nucleus, the atomic number changes from Z to $Z + 1$. Generally, β^- decay occurs for the nuclide having excess neutrons other than the stable isotopes. For example,

$$^{42}_{18}\mathrm{Ar} \rightarrow {}^{42}_{19}\mathrm{K} + e^- + \bar{\nu}_e \tag{5.11}$$

The Q value for the β^- decay is given by

$$Q = [M(A,Z) - Zm_e] - [M(A,Z+1) - (Z+1)m_e + m_e]$$

$$= M(A,Z) - M(A,Z+1) \tag{5.12}$$

in which M represents the atomic mass and m_e the electron mass. This energy is shared by a β-particle and antineutrino.

$$E_{\beta^-} + E_{\bar{\nu}} = Q \tag{5.13}$$

Both E_{β^-} and $E_{\bar{\nu}}$ distribute from 0 to Q. Therefore, the energy spectrum of β-particle becomes a continuous spectrum. According to Fermi's theory, the energy spectrum of β-particles is given by

$$\frac{dn}{dT} \propto (Q-T)^2(T+mc^2)\sqrt{T(T+2mc^2)} \tag{5.14}$$

where dn = the relative intensity, T = the kinetic energy, and m = the mass of β-particle. Figure 5.3 shows the energy spectrum for the β^- rays from ^{40}K decay with $Q = 1{,}314$ keV (100%). If a single Q, then the average energy is $\sim Q/3$. For ^{40}K, the average energy of β^- particles is 536 keV.

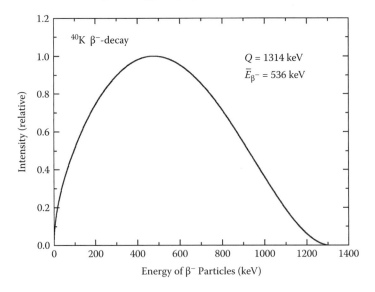

FIGURE 5.3
Energy spectrum of β^- particles ejected by β^- decay of ^{40}K.

EXAMPLE 5.2

Calculate the value for the decay of Co-60.

SOLUTION 5.2

The decay of Co-60 is

$$^{60}_{27}\text{Co} \rightarrow {}^{60}_{28}\text{Ni} + e^- + \bar{\nu}_e$$

The antineutrino mass can be neglected. Therefore,

$$Q = M\left({}^{60}_{27}\text{Co}\right) - M\left({}^{60}_{28}\text{Ni}\right)$$

$$= (59.93382 - 59.93079) \text{ u}$$

$$= 0.00303 \text{ u} = 0.00303 \cdot 931.5 \text{ MeV} = 2.82 \text{ MeV}$$

The nickel-60 nucleus is in an excited state; therefore, it will emit a γ-ray to attain its ground state.

5.1.3 γ-Decay

One or several γ photons are emitted from the excited states of the daughter nucleus produced by the α- or β-decay. In the γ-transition, Z and A both do not change. The γ-ray spectrum is discrete and specific to the nuclide. The lifetime of the nuclear excited states is usually $\sim 10^{-10}$ s; therefore, γ-rays are promptly emitted after the decay. However, there are some exceptions due to the selection rule for photon emission. The half-life of the excited state of $^{137}_{56}\text{Ba}$ produced by the decay of $^{137}_{55}\text{Cs}$ is 2.55 minutes. Such a long life state is called a metastable state. The relevant state is denoted by $^{137\text{m}}_{56}\text{Ba}$. Another example of the metastable nuclide is $^{99\text{m}}_{43}\text{Tc}$ produced by the β⁻ decay of $^{99}_{42}\text{Mo}$. This state with the half-life of 6 h causes the isomeric transition (IT) to the ground state.

$$^{99\text{m}}_{43}\text{Tc} \rightarrow {}^{99}_{43}\text{Tc} + \gamma \tag{5.15}$$

The energy of γ-ray is equal to the difference between the metastable state and the ground state. Figure 5.4 shows the decay scheme of $^{99}_{42}\text{Mo}$.

5.1.4 Internal Conversion

In internal conversion an electron is ejected from the atom by transferring the energy of the excited state to an electron in the K- or L-shell of the atom. This is regarded as an alternative to photon emission. The kinetic energy of ejected electron E_e, or E_{IC}, becomes $E_\gamma - E_B$, in which E_B is the binding energy The process of internal conversion occurs in competition with γ-decay. The relative proportion of the two processes is given by the total conversion

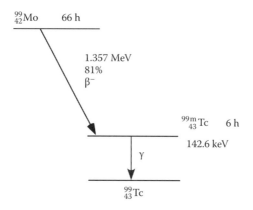

FIGURE 5.4
Decay scheme of ^{99}Mo (Lederer et al. 1978).

coefficient, α. The internal conversion coefficient α is defined by the ratio of the number of conversion electrons N_e to that of competing photons N_γ.

$$\alpha = N_e/N_\gamma$$

$$= \alpha_K + \alpha_{L1} + \alpha_{L2} + \cdots \tag{5.16}$$

where α_K, α_{L1}, and α_{L2} are the partial conversion coefficients for the individual subshells. The magnitude of a vacancy cascade depends on the individual subshells. The partial conversion coefficients are given by Rosel et al. (1978), Coursol et al. (2000), and Gorozhankin et al. (2002). For ^{137}Cs, 95% of β-decay goes to the metastable state lying at 662 keV of ^{137}Ba. The relative intensity of the γ-rays emitted at this time is 85%. Internal conversion accounts for the 10% difference. The energies of conversion electrons for the K-shell and L-shell of ^{137}Ba are 624 keV and 656 keV, respectively. These are the line spectrum. Of particular interest is the Auger electron emitting the radioisotopes ^{123}I, ^{124}I, and ^{125}I (Kassis 2004; Nikjoo et al. 2008).

5.1.5 β^+ Decay

β^+ decay is a phenomenon that a proton changes into a neutron following the emission of a positron (β^+) and a neutrino.

$$p \rightarrow n + e^+ + \nu_e \tag{5.17}$$

When this decay occurs within a nucleus, the atomic number changes from Z to Z – 1. Generally, β^+ decay occurs for the nuclide having excess protons other than the stable isotopes. For example,

$$^{18}_{9}F \rightarrow {}^{18}_{8}O + e^+ + \nu_e \tag{5.18}$$

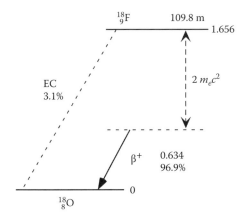

FIGURE 5.5
Decay scheme of ^{18}F (Lederer et al. 1978).

Figure 5.5 shows the decay scheme of $^{18}_{9}$F. The Q value for the β^+ decay is given by

$$Q = [M(A,Z) - Zm_e] - [M(A,Z-1) - (Z-1)m_e + m_e]$$
$$= M(A,Z) - M(A,Z-1) - 2m_e \tag{5.19}$$

in which M represents the atomic mass and m_e the electron mass. In order to result in positron emission, therefore, the mass of the parent atom must be greater than that of the daughter atom by at least $2mc^2 = 1.022$ MeV. The released energy Q is used for the kinetic energy of the positron and neutrino. The nuclide that emits positrons is called the positron emitter. The peculiar phenomenon of the positron emitter in matter is to accompany two annihilation photons of 511 keV. They are in opposite directions of each other.

5.1.6 Electron Capture

Electron capture (EC) is a phenomenon that a proton changes into a neutron by capturing an orbital electron following the emission of a neutrino and characteristic x-ray.

$$p + e^- \rightarrow n + \nu_e \tag{5.20}$$

When this decay occurs within a nucleus, the atomic number changes from Z to $Z - 1$. Generally, EC occurs for the nuclide having excess protons other than the stable isotopes. EC competes with β^+ decay, which results in the same change. For example,

$$^{18}_{9}F + e^- \rightarrow ^{18}_{8}O + \nu_e \tag{5.21}$$

The Q value for EC is given by

$$Q = \left[M(A,Z) - Zm_e + m_e - B\right] - \left[M(A,Z-1) - (Z-1)m_e\right]$$
$$= M(A,Z) - M(A,Z-1) - B \qquad (5.22)$$

where B is the mass equivalent to the binding energy of the atomic shell. In order for the EC to occur, therefore, the mass of the parent atom must be greater than that of the daughter atom by the binding energy. The orbital electron in the K-shell is usually captured. The vacancy is necessarily generated in the inner shell; therefore, the characteristic x-ray of the daughter nucleus is always emitted. EC is detected by observation of the characteristic x-rays and the Auger electrons. In calculations related to Auger electron cascades the relative proportion of vacancies created by EC in the individual subshells representing the distribution of initial vacancies in the decaying isotope is needed. The electron capture probabilities for i- and j-shells is given, for example, for subshells K and L_1 by

$$\frac{P_{L1}}{P_K} = \frac{n_{L1} q_{L1}^2 B_{L1} \beta_{L1}^2}{n_K q_K^2 B_K \beta_K^2} \qquad (5.23)$$

and

$$\frac{q_{L1}}{q_K} = \frac{E_{EC} - E_{L1}}{E_{EC} - E_K} \qquad (5.24)$$

For radionuclide decay by EC the sum of probabilities for EC in each individual subshell is equal to unity:

$$P_{EC} = P_K + P_{L1} + P_{L2} + P_{L3} + \ldots = 1 \qquad (5.25)$$

EXAMPLE 5.3

Calculate the difference between the values for the positron emission of ^{18}F and electron capture by the same nucleus.

SOLUTION 5.3

For positron emission (Equation (5.18)),

$$Q = M\left(^{18}_9F\right) - M\left(^{18}_8O\right) - 2m_e$$

$$= (18.000938 - 17.99916 - 2 \cdot 5.4858 \cdot 10^{-4})\,u$$

$$= 6.8084 \cdot 10^{-4}\,u = 6.8084 \cdot 10^{-4} \cdot 931.5\,\text{MeV} = 0.634\,\text{MeV}$$

On the other hand, for electron capture (Equation (5.21)), the value is calculated from

$$Q = M\left(^{18}_{9}\text{F}\right) - M\left(^{18}_{8}\text{O}\right) - B$$

$$= (18.000938 - 17.99916)\,\text{u-696.7 eV}$$

$$= 0.001778 \cdot 931.5\,\text{MeV-6.967} \cdot 10^{-6}\,\text{MeV} = 1.6562\,\text{MeV}$$

where is the K-shell electron binding energy of fluorine (tabulated in, for instance, Cardona and Ley 1978).

The difference in the values obtained from both processes is 1.022 MeV (which is about twice the electron mass energy equivalent; in this case, the binding energy of the K-shell electron is much lower than the nuclear mass difference).

5.1.7 Radiative and Nonradiative Transitions

Absorption of photons through the photoelectric effect or the interaction of ions with the target results in ionization and excitation and creation of vacancy in the inner shells of the atom. Subsequently, the vacancy may be filled by radiative (x-ray) or nonradiative (Auger) processes. Electron vacancies are generated when an atom is ionized by an incident electron beam and electrons are knocked out from their shells. This process is called inner shell ionization. When an electron vacancy is created, the atom is said to be in an excited state. The atom in such an excited state returns to the ground state by filling the electron vacancy by an electron from one of the atomic subshells. For example, if the K-shell is excited, the Auger electron may originate in the L-shell. The second way of creating a vacancy is by internal conversion (IC). In internal conversion the excitation energy of the nucleus is transferred to an atomic electron, which is subsequently emitted with the energy $E_g - E_b$, where E_g is the excitation energy and E_b the binding energy of the electron in its atomic shell. And the third way of generating electron vacancy is by electron capture (EC). In the process of electron capture the orbital electron of an atom interacts with a nucleus proton to form a neutron, and a neutrino is emitted. The probability of such an interaction depends on the overlap between the atomic electron and the nucleus. This is highest for the K-shell. If the filling of the vacancy is accompanied by the emission of X-photon, it is said to be a radiative transition. If the filling of the vacancy is accompanied by the emission of an electron, it is a nonradiative transition. Above the K-shell, if the filling electron originates from the same family of subshells, it is called a Coster-Kronig transition.

Figure 5.6 shows various scenarios leading to the emission of photons or electrons from an atom when in an excited state. The relative probability of Auger emission and photon emission is measured by the fluorescence yield

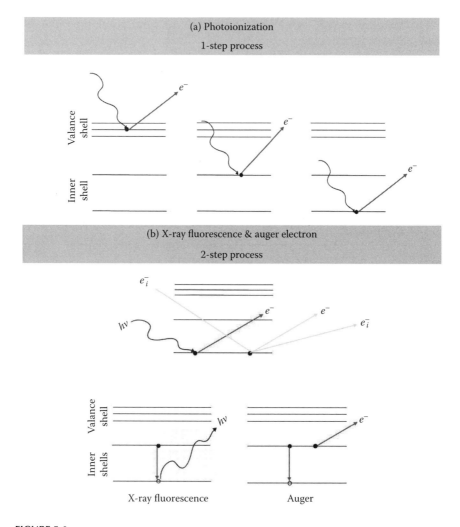

FIGURE 5.6
Various scenarios leading to the emission of photons or electrons from an atom when in an excited state.

such that $\omega + a + f = 1$, where a and f refer to Auger and Coster-Kronig yields. Fluorescence yield is the relative probability of Auger emission and x-ray such that, for the K-shell, ω is the number of K-photons/number of K-shell vacancies. For the heavy atoms x-radiation is the more probable because of the high nuclear charge. For light atoms the Auger effect is predominant. Other processes that may be of importance are super Coster-Kronig transitions, autoionization, double Auger effect, Auger-Coster-Kronig transitions, interatomic Auger effect, electron shake-up and shake-off, and plasma excitations.

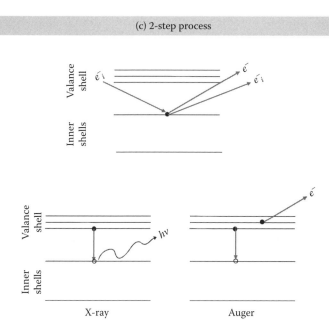

FIGURE 5.6
(Continued)

EXAMPLE 5.4

Calculate the Auger electron energy following the K-shell ionization of water vapor.

SOLUTION 5.4

The K-shell binding energy (B_K) of water vapor is 539.7 eV, and the first ionization energy (B_1) is 12.62 eV. Therefore, the Auger electron energy is

$$E_e = B_K - 2 \cdot B_1 = (539.7 - 2 \cdot 12.62) \, \text{eV} = 514.46 \, \text{eV}$$

5.2 Formulas of Radioactive Decay

5.2.1 Attenuation Law

The disintegration rate of radioactive nuclides is represented by activity. The unit of activity Bq is defined by the number of disintegrated atoms per second. The activity of a radioisotope decreases exponentially in time.

Assuming the number of radioactive nuclides in a sample N, the variation of the number dN in time dt is given by

$$dN = -\lambda N dt \tag{5.26}$$

where λ is a constant specific to a radioactive nuclide called the decay constant. The activity A becomes

$$A = \frac{-dN}{dt} = \lambda N \tag{5.27}$$

Solving Equation (5.26),

$$N = N_0 e^{-\lambda t} \tag{5.28}$$

is obtained, in which N_0 is the number at $t = 0$. The same form can be applied to the activity A. Figure 5.7 shows the exponential decay pattern for the activity as a function of the time. The time at which the activity reduces by one-half is called the half-life. The relationship between the half-life T and λ becomes

$$T = \frac{\ln 2}{\lambda} = \frac{0.693}{\lambda} \tag{5.29}$$

The attenuation law is written by

$$\frac{A}{A_0} = e^{-0.693t/T} = \left(\frac{1}{2}\right)^{t/T} \tag{5.30}$$

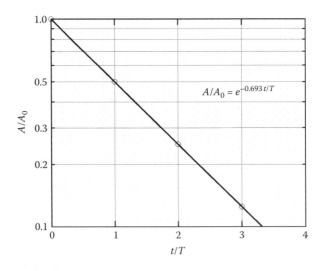

FIGURE 5.7
Exponential decay of radioactivity.

EXAMPLE 5.5

^{60}Co source is a γ-emitter, which is used in medicine and industry. The half-life of the ^{60}Co source is 5.3 years. In order to account for the source strength decrease, a correction factor could be applied. Calculate the source strength decrease each year.

SOLUTION 5.5

$$\frac{A}{A_0} = \left(\frac{1}{2}\right)^{t/T} = \left(\frac{1}{2}\right)^{1/5.3} = 0.877$$

The activity of the source after a year is 87% of the initial activity.

5.2.2 Specific Activity

The specific activity of a sample is defined by the activity per unit mass. The specific activity S is determined by T and atomic mass M. Because the number of atoms per gram is $6.022 \times 10^{23}/M$, Equation (5.27) is changed into

$$S = \frac{4.17 \times 10^{23}}{MT} \tag{5.31}$$

The unit of S is Bq g^{-1} if T s.

EXAMPLE 5.6

What is the specific activity of ^{60}Co and ^{226}Ra?

SOLUTION 5.6

$$S(^{60}\text{Co}) = \frac{4.17 \times 10^{23}}{MT} = \frac{4.17 \times 10^{23}}{60 \times 5.4 \times 365 \times 24 \times 3600} = 4.08 \times 10^{13}$$

The specific activity of ^{226}Ra is 1 Ci/g or 3.7×10^{10} Bq/g.

5.2.3 Radioactive Equilibrium

5.2.3.1 Secular Equilibrium

A parent nucleus with the half-life of T_1 decays to a daughter nucleus with the half-life of T_2. It is assumed that T_1 is much longer than T_2, namely, $T_1 \gg T_2$. The parent activity A_1 is regarded as constant in the short time. The variation of the number of daughter atoms is given by

$$\frac{dN_2}{dt} = A_1 - \lambda_2 N_2 \tag{5.32}$$

The solution A_2 is obtained putting the initial condition for the daughter as N_{20} at $t = 0$.

$$A_2 = A_1(1 - e^{-\lambda_2 t}) + A_{20}e^{-\lambda_2 t} \tag{5.33}$$

If starting from the pure parent, $A_{20} = 0$ at $t = 0$. This condition is called the secular equilibrium.

5.2.3.2 General Formula

When the relative magnitudes of T_1 and T_2 are unrestricted, the differential equation for N_1 and N_2 becomes

$$\frac{dN_2}{dt} = \lambda_1 N_1 - \lambda_2 N_2 \tag{5.34}$$

The first term of the right side is

$$\lambda_1 N_1 = \lambda_1 N_{10}e^{-\lambda_1 t} \tag{5.35}$$

Therefore, Equation (5.34) is changed into

$$\frac{dN_2}{dt} + \lambda_2 N_2 = \lambda_1 N_{10}e^{-\lambda_1 t} \tag{5.36}$$

The solution of this equation is obtained assuming the initial condition $N_{20} = 0$ at $t = 0$.

$$N_2 = \frac{\lambda_1 N_{10}}{\lambda_2 - \lambda_1}\left(e^{-\lambda_1 t} - e^{-\lambda_2 t}\right) \tag{5.37}$$

EXAMPLE 5.7

Calculate the activity of ^{222}Rn from a 1 g ^{226}Ra sample after a week (half-life of ^{226}Ra = 1,600 years, half-life of ^{222}Rn = 3.8 days).

SOLUTION 5.7

$$A_{Rn} = A_{Ra}\left(1 - e^{-\lambda_{Rn}t}\right) = 3.7 \times 10^{10}(\text{Bq/g}) \times 1(\text{g}) \times \left(1 - e^{-\frac{0.693}{3.8} \times 7}\right) = 2.67 \times 10^{10}\,\text{Bq}$$

5.2.3.3 Transient Equilibrium

For the case of $T_1 \geq T_2$ and $N_{20} = 0$, $A_2 = \lambda_2 N_2$ increases slowly. The second term in the bracket can be neglected in comparison with the first term as time goes by.

$$\lambda_2 N_2 = \frac{\lambda_1 \lambda_2 N_{10}e^{-\lambda_1 t}}{\lambda_2 - \lambda_1} \tag{5.38}$$

Because $A_1 = \lambda_1 N_{10} e^{-\lambda_1 t}$ is the activity of the parent, the above equation is written as

$$A_2 = \frac{\lambda_2 A_1}{\lambda_2 - \lambda_1} \tag{5.39}$$

This condition is called the transient equilibrium.

5.2.3.4 Nonequilibrium

For the case of $T_1 < T_2$, the activity of the daughter nucleus becomes maximum and attenuates. The parent soon decays and the daughter remains. The equilibrium is not realized.

5.3 Summary

1. The process where a nucleus transfers to more stable states is called the disintegration of the nucleus, in which excess energies are emitted as various types of radiation.

2. Emitted radiations following the decay are α, β^-, β^+, γ, internal conversion electrons, and so on. Electron capture is detected by observing the characteristic x-rays and Auger electrons.

3. Activity in Bq units attenuates exponentially as a function of time. The time when the activity reduces to half is called the half-life.

4. Half-lives of the parent nucleus and the daughter nucleus are assumed to be T_1 and T_2, respectively. If $T_1 \gg T_2$, then the secular equilibrium is realized. If $T_1 \geq T_2$, then the transient equilibrium is realized.

QUESTIONS

1. Write the decay of tritium. How much of 10 g tritium is left after 25 years?

2. A luminous watch dial contains 5 microcuries of the isotope radium. How many decays per second are taking place?

3. A fresh sample of ^{210}Bi weighs 1 pg and decays to ^{210}Po by emitting a β^- particle with a $T_{1/2}$ of 5 days. ^{210}Po decays to ^{205}Pb by emitting an α-particle with a $T_{1/2}$ of 140 days. Calculate the maximum mass of ^{210}Po present at any time and the activity.

4. If two radioactive nuclei A and B, produced in the process of nuclear fission, are characterized by the disintegration constants λ_1 and λ_2, and if the probability that a time less than T elapses between the subsequent disintegrations of A and B is represented by $W(T)$, show that $W(T) = 1 - (\lambda_1 + \lambda_2)^{-1} (\lambda_1 e^{-\lambda_2 T} + \lambda_2 e^{-\lambda_1 T})$.

References

Cardona M, Ley L. 1978. *Photoemission in Solids. I. General Principles.* Berlin: Springer-Verlag. With additional corrections.

Coursol N, Gorozhankin VM, Yakushev EA, Briancon C, Vylov T. 2000. Analysis of internal conversion coefficients. *Appl. Radiat. Isot.* 52(3): 557–567.

Gorozhankin VM, Coursol N, Yakushev EA, Vylov Ts, Briançon C. 2002. New features of the IC(4) code and comparison of internal conversion coefficient calculations. *Appl. Radiat. Isot.* 56(1–2): 189–197.

Kassis AI. 2004. The amazing world of Auger electrons. *Int. J. Radiat. Biol.* 80(11–12): 789–803.

Lederer CM, Shirley VS. 1978. *Table of Isotopes*, 7th ed. New York: John Wiley & Sons.

Nikjoo H, Emfietzoglou D, Charlton DE. 2008. The Auger effect in physical and biological research. *Int. J. Radiat. Biol.* 84(12): 1011–1026.

Rosel F, Fries HM, Alder K, Pauli HC. 1978. Internal conversion coefficients for all atomic shells. *At. Data Nucl. Data Tables* 21: 91 ± 289.

For Further Reading

Turner JE. 1995. Atoms, Radiation, and Radiation Protection, 2nd ed. New York: John Wiley & Sons.

6

X-Rays

6.1 Generation of X-Rays

In 1895, German physicist W.C. Roentgen discovered a new type of radiation in the course of an experiment using the Crookes' vacuum tube. This radiation penetrated opaque materials such as black paper or wood. He named it x-ray and clarified its properties such as penetrability, fluorescence action, imaging ability, and ionization. The discovery of x-rays is the starting point of the investigation on ionizing radiation and its application to various fields.

X-rays are produced when electrons are incident on target. The electron collides with a target atom and loses a major part of its energy by ionization and excitation of an atom. In addition, the electron decelerated in the field of an atomic nucleus loses its energy by emitting the x-ray photons. An electron having the kinetic energy T can produce photons having the energy distributed less than T.

$$h\nu \leq T \tag{6.1}$$

As a result, monochromatic electron beams generate a continuous energy spectrum with the maximum energy equal to the beam energy. These continuous x-rays are called bremsstrahlung or braking radiation. Bremsstrahlung is likened to photons shaken off an electron that suffered from a sudden brake in the nuclear field.

Figure 6.1 shows the schematic diagram of an x-ray generator. This structure is called the Coolidge tube and consists of the cathode and anode. The atmospheric pressure is kept at less than 10^{-6} mmHg. Thermal electrons emitted from the cathode are accelerated and collide with the anode target. Only a few percent of electron kinetic energy is converted to x-rays. The major part of the remaining energy is converted to heat. The x-ray tube is usually used under the voltage less than 300 kV. Only 1% of the electron energy is converted to useful x-rays, and 99% is converted to heat and warms the anode. Therefore, tungsten having a high melting point is frequently used as the target. In addition, the structure of the anode is devised in such a way as to decrease the temperature.

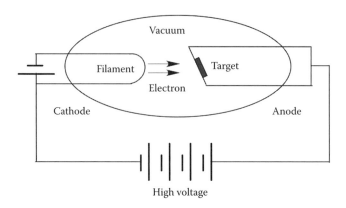

FIGURE 6.1
Schematic diagram of x-rays generator.

For high-energy electrons, acceleration of electrons is made by magnetic field or microwave used in a betatron or linac. The higher the electron energy, the higher the generation efficiency that is achieved. The angular distributions of x-rays vary. Bremsstrahlung is emitted in all directions. That is a complex function of the energy of electrons and bremsstrahlung photons. Segre gave the average emission angle of x-rays as

$$\bar{\theta} \approx \frac{1}{1 + T/mc^2} \quad \text{(radian)} \tag{6.2}$$

in which T is the kinetic energy of electrons, and m is the electron rest mass. Therefore, the dominant angle is the forward direction for high-energy electrons and around 60° for low-energy electrons. The extraction direction is different between the x-ray tube and high-energy linac.

EXAMPLE 6.1

Common energy for x-ray tubes is about 120 kVp, and for radiotherapy accelerators is about 6 MV. What is the most probable emission angle for the two examples in order to design the target angle?

SOLUTION 6.1

$$\bar{\theta} \approx \frac{1}{1 + T/mc^2} =$$

$$\frac{1}{1 + 120/511} = 0.809 (\text{radian}) = 46.4°$$

$$\frac{1}{1 + 6000/511} = 0.078 (\text{radian}) = 4.5°$$

6.2 Continuous X-Rays

The x-ray intensity rapidly increases as the tube voltage increases if the current is kept constant. The wavelength of the photon with the maximum energy becomes the shortest. Using the relationship, $\lambda[nm] = 1.24/h\nu[keV]$, the minimum wavelength for the tube voltage 50 kV becomes $\lambda_{min} = 1.24/50 = 0.0248$ nm $= 0.248$ Å. Generally, the energy of x-rays is referred to by the peak voltage of the x-ray tube, kV_p. The maximum voltage V_{max} in kV units and the minimum wavelength λ_{min} in $\overset{\circ}{A}(= 10^{-8} cm)$ units is related by the Duane-Hunt's equation:

$$V_{max} \times \lambda_{min} = 12.4 \qquad (6.3)$$

At the present it is rather convenient to represent the x-ray spectrum as a function of energy unit eV in place of wavelength. Figure 6.2 shows the measured energy spectrum extracted from a diagnostic x-ray of 80 kV$_p$. Figure 6.3 shows the energy spectrum for the therapeutic x-rays generated by 4 MV linac. Naturally the number of bremsstrahlung photons is smaller as the energy is lower. The reason is that the low-energy component is reduced because of the absorption by the metal filter and the target itself equipped in the apparatus. The energy spectrum not extracted from the x-ray tube but produced at the target is different. Detailed descriptions for the Koch-Motz's bremsstrahlung cross sections and the thick target spectrum are given in Section 12.5.

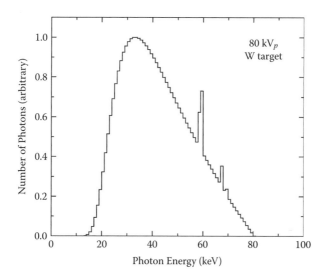

FIGURE 6.2
Energy spectrum of x-rays generated by the tube voltage of 80 kV.

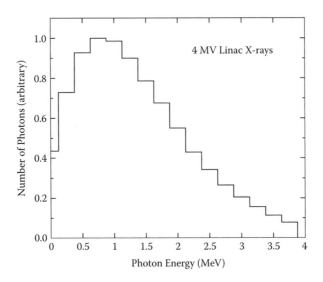

FIGURE 6.3
Energy spectrum of x-rays generated by 4 MV linac.

6.3 Characteristic X-Rays

The binding energy of tungsten K-shell is $E_K = 69.525$ keV. If the tube voltage is higher than that, the projectile ejects an orbital electron of the target atom. Then, discrete or discontinuous x-rays are generated. These x-rays are emitted when the electrons of the outer orbit fill the vacancy of the inner shell. These are called the characteristic x-rays because those are characteristic of the target element. Figure 6.2 shows the characteristic x-rays as the lines overlying on the continuous bremsstrahlung spectrum. The K-shell vacancy is filled by the electron from L-shell and M-shell, etc., and the x-rays are marked by K_α and K_β, etc., respectively. For the L-shell vacancy, L_α and L_β x-rays are emitted. The orbital energies other than K-shell do not degenerate; namely, those differ a little with each other. Therefore, K x-ray has a fine structure. L-shell of tungsten consists of three subshells with the electron binding energies of LI = 12.098, LII = 11.541, and LIII = 10.204 keV. The transition LIII $\rightarrow$ K produces $K_{\alpha 1}$ x-ray with the energy of E_K-E_{LIII} = 69.525 − 10.204 = 59.321 keV. The transition LI $\rightarrow$ K produces $K_{\alpha 2}$ x-ray with the energy of 57.984 keV. The frequency of K_α x-ray is related to the atomic number by

$$\sqrt{v} = k(Z - S) \tag{6.4}$$

where Z = atomic number, and k and S are constants. This is called Moseley's law. This law played an important role in understanding the atom structure and the periodic table.

EXAMPLE 6.2

The binding energies of innermost shell electrons for molybdenum are as follows:

$$K = 20{,}000 \text{ eV, LI} = 2{,}866 \text{ eV, LII} = 2{,}625, \text{ and LIII} = 2{,}520 \text{ eV}$$

Calculate the characteristic x-ray $K_{\alpha 1}$, $K_{\alpha 2}$, $K_{\alpha 3}$ for this target.

SOLUTION 6.2

$$K_{\alpha 1}: L_{III} \rightarrow K: \ E_K - E_{LIII} = (20000 - 2520) \text{ eV} = 17.48 \text{ keV}$$

$$K_{\alpha 2}: L_{II} \rightarrow K: \ E_K - E_{LII} = (20000 - 2625) \text{ eV} = 17.375 \text{ keV}$$

$$K_{\alpha 3}: L_{I} \rightarrow K: \ E_K - E_{LI} = (20000 - 2866) \text{ eV} = 17.134 \text{ keV}$$

6.4 Auger Electrons

Even if an L electron of an atom moves to the vacancy of the K-shell, a photon is not necessarily emitted. This phenomenon frequently occurs for a small Z element. There is a probability of an alternative nonoptical transition in which an L electron is ejected from an atom. In consequence, two vacancies are generated in the L-shell. Such an ejected electron is called the Auger electron. Figure 6.4 illustrates the emission of an Auger electron on a water molecule. The downward arrow shows the transition of the electron from the $1b_1$ level to the oxygen-K-shell vacancy. Then, the energy of $E_{OK} - E_{1b1}$ is released. In place of a photon emission, this energy is given to another electron on the $1b_1$ orbit. This electron is ejected from an atom with the kinetic energy of

$$T = E_{OK} - E_{1b1} - E_{1b1} \tag{6.5}$$

For a water molecule with $E_{OK} = 539.7$ eV and $E_{1b1} = 12.62$ eV, the energy of the Auger electron becomes 514.5 eV.

The K fluorescence yield is defined by the number of KX-photon emission per K-shell vacancy. The yield Y_K as a function of Z is approximated by

$$Y_K = \frac{1}{1 + (33.6/Z)^{3.5}} \tag{6.6}$$

The yield for smaller Z is nearly 0, and that for larger Z nearly 1. On the contrary, emission of Auger electrons occurs frequently for the smaller Z in comparison with photon emission. The inner shell vacancy as the source of Auger electron emission is generated not only by electron collision but also by orbital electron capture, internal conversion, and photoelectric effect. The emission of an Auger electron increases one vacancy of atomic shell. The inner

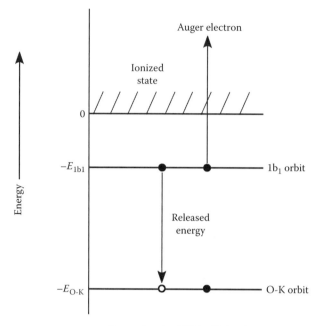

Energy Levels of Water Molecule

FIGURE 6.4
Schematic diagram of Auger electron emission on water molecule.

shell vacancy is occupied by the Auger electron and a simultaneously weakly bound electron is ejected. Therefore, the Auger cascade occurs for a relatively heavy atom. An ion at first with only one inner shell vacancy changes to a highly charged ion by Auger cascade. This phenomenon is applied to the Auger therapy in radiology. An Auger emitter such as ^{125}I is taken in DNA or biological molecules. [125]I decays by electron capture. Auger cascades release about 20 electrons, which deposit plenty of energy (~1 keV) within a few nm. Highly charged [125]Te remains. As a result, biological effects such as DNA strand break, chromosome aberration, and cell death are induced.

6.5 Synchrotron Radiation

A phenomenon that was theoretically well known was that an electromagnetic wave is emitted when a charged particle moves with acceleration. In 1947, this was observed using electron synchrotron. Such an electromagnetic wave was named synchrotron radiation (SR) or synchrotron light. When high-energy relativistic electrons are forced to travel in a circular path by a magnetic field, synchrotron radiation is produced. If the speed of an electron

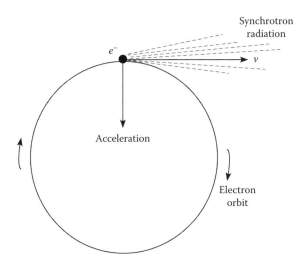

FIGURE 6.5
Schematic diagram of synchrotron radiation.

is near the speed of light, photons are emitted to the narrow angular region around the electron direction—i.e., the tangential direction. Figure 6.5 shows the scheme of synchrotron radiation. The radiation produced may range over the entire electromagnetic spectrum, from radio waves to infrared light, visible light, ultraviolet light, x-rays, and γ-rays. According to the detailed calculation, the intensity SR per unit time and unit wavelength is given by

$$I(\lambda,t) = 7.51 \times 10^{-8} \frac{E^7}{R^3} G(y) \quad \text{Js}^{-1}\text{Å}^{-1} \tag{6.7}$$

in which E (GeV) is the kinetic energy of electrons, R (m) is the orbital radius, and $G(y)$ is a function providing the spectrum shape. Figure 6.6 shows the SR spectra for various electron energies. The total radiant energy per unit time is given by

$$I(t) = 6.77 \times 10^{-7} \frac{E^4}{R^2} \quad \text{Js}^{-1} \tag{6.8}$$

The energy lost per turn is approximately

$$\Delta E = 88.5 \frac{E^4}{R} \quad \text{keV} \quad \text{turn}^{-1} \tag{6.9}$$

ΔE becomes 5.53 MeV for $E = 5$ GeV and $R = 10$ m. Therefore, acceleration is needed to compensate the energy lost for large E and small R. The SR generated by an electron synchrotron or storage ring is a powerful light source covering a broad range of wavelength. It has characteristics superior to the conventional light sources: (1) a high-intensity continuous spectrum, (2) a good directivity, (3) a polarized light, and (4) a short time pulse.

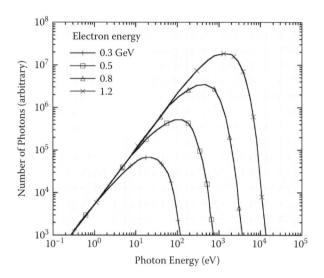

FIGURE 6.6
Energy spectra of synchrotron radiation for various electron energies.

EXAMPLE 6.3

The European Synchrotron Radiation Facility (ESRF) is one of the synchrotron research facilities in Europe. The electron energy in the synchrotron storage ring is 6 GeV, and the storage ring is a semicircle with 844.4 m circumference. Calculate the total radiant energy per unit time and the energy lost per turn in the storage ring.

SOLUTION 6.3

$$I(t) = 6.77 \times 10^{-7} \frac{E^4}{R^2} = 6.77 \times 10^{-7} \frac{6^4}{(844.4/2\pi)^2} = 4.9 \times 10^{-8} \, \text{Js}^{-1}$$

$$\Delta E = 88.5 \frac{E^4}{R} = 88.5 \frac{6^4}{844.4/2\pi} = 853.5 \quad \text{keV} \quad \text{turn}^{-1}$$

6.6 Diffraction by Crystal

The penetrability of x-rays is used for various fields of application. The structure analysis of crystal uses the wave motion of x-rays. A solid forms a well-ordered crystal, unlike vapor or liquid. If solid atoms or solid molecules are spatially arranged in a regular manner, this is called the crystal lattice.

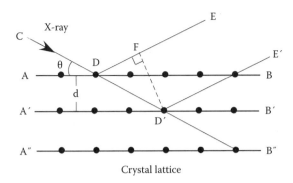

FIGURE 6.7
Bragg diffraction by crystal.

Each point in the crystal lattice is called the point of a lattice. The distance between the points of a lattice of a solid crystal is ordinarily a few nm. X-rays are used as a rule to measure the crystal structure because of the similarity of wavelength. When x-rays irradiate crystal, a major part of x-rays passes on the straight line. They are penetrated x-rays. A part of x-rays scattered by an atom becomes the diffracted x-rays enhancing each other at the specific direction. This phenomenon is similar to the diffraction of visible light impinging the diffraction lattice.

Assuming a plane in a crystal, there is a case that a lot of atoms are arranged in a regular manner on the plane. Such a plane is called the plane of atomic arrangement. Figure 6.7 shows the planes parallel forming with an equal distance d, which is called the lattice constant. It is assumed that x-rays with the wavelength λ impinge with the angle θ vs. AB from the direction CD and reflect the direction DE. A part of the remaining x-rays is reflected on the planes A′B′ and A″B″ and goes to D′E′ and D″E″ parallel to DE. The difference of path length of the reflected x-rays between a plane and the next one is equal to DD′ – DF. Using

$$DD' = \frac{d}{\sin\theta}, \quad DF = \frac{d}{\sin\theta}\cos 2\theta \qquad (6.10)$$

then

$$DD' - DF = \frac{d}{\sin\theta}(1 - \cos 2\theta) = 2d\sin\theta \qquad (6.11)$$

is obtained. If this value is equal to the integer times the wavelength λ,

$$2d\sin\theta = n\lambda \quad (n = 1,2,3,\ldots\ldots) \qquad (6.12)$$

is held and the reflected wave is enhanced by interference with each other. This is called the Bragg's reflection condition. The integer n is the order of reflection. For known λ, the measurement of the lattice constant d is possible, while the wavelength of x-rays can be determined by measuring the position of *moire* (denoted by θ) if the lattice constant of crystal is known.

EXAMPLE 6.4

The d-spacing of a diamond is 1.075 Å. At which angle θ do we expect to detect the first-order interference pattern, given the impinging light has a wavelength of 1.54 Å? What is the maximum wavelength of light needed for detecting the interference pattern at the same angle for a crystal with the lattice constant of 1.261 Å?

SOLUTION 6.4

$$\sin(\theta) = \frac{1.54}{2 \cdot 1.075} = 0.716 \Rightarrow \theta = 45.75°$$

For the d-spacing of 1.261 Å and the angle $\theta = 45.75°$, the maximum wavelength is

$$\lambda_{max} = 2 \cdot 1.261 \text{ Å} \cdot 0.716 = 1.806 \text{ Å}$$

6.7 Summary

1. When electrons hit the target material, electrons suffer a sharp breaking near the nucleus and generate x-rays.

2. Monochromatic electron beam generates the continuous x-ray spectrum whose maximum is the beam energy. This continuous x-ray is called bremsstrahlung.

3. The relationship $V_{max} \times \lambda_{min} = 12.4$ holds for the peak voltage V_{max} (kV) and the minimum wavelength λ_{min} (Å).

4. Electrons eject the orbital electron from the target atom. Either characteristic x-ray or Auger electron is emitted when the electrons of the outer shell occupy the inner shell vacancy.

5. High-speed electrons moving in circular orbit emit the synchrotron radiation toward the tangential direction. Synchrotron radiation is useful as the polychromatic light source, similar to bremsstrahlung.

QUESTIONS

1. Calculate a photon energy generated when an electron with the initial velocity of 0.8*c* is braked to 0.6*c* by the nuclear electric field.
2. The efficiency of bremsstrahlung generation for diagnostic x-ray tube is represented by $\eta = kZV$, in which k is a constant, V is the tube voltage [kV] and Z is the atomic number of target. On an x-ray generator with W target, η is 0.74% for 100 kV. What is the bremsstrahlung output in Watt units for 80 kV and the tube current 200 mA.
3. An x-ray beam generated by 150 kVp is hardened by a tin filter. What will be the best choices of a second and a third filter to stop fluorescent radiation?

For Further Reading

Birch R, Marshall M, Ardran GM. 1979. *Catalogue of Spectral Data for Diagnostic X-Rays*. London: Hospital Physicists' Association.

Johns HE, Cunningham JR. 1974. *The Physics of Radiology*, 3rd ed. Springfield, IL: Charles C. Thomas Publisher.

Turner JE. 1995. *Atoms, Radiation, and Radiation Protection*, 2nd ed. New York: John Wiley & Sons.

7

Interaction of Photons with Matter

7.1 Types of Interaction

Energetic charged particles when moving steadily lose energy by the electric interactions with the atoms in matter. On the other hand, photons lose energy by a different manner than the electric interaction because of their charge neutralities. Namely, photons can go forward a certain distance before an interaction with an atom occurs. That distance is statistically dominated by the interaction probability per unit length, which depends on the medium and the photon energy. Sometimes interacted photons are absorbed and disappear, or are scattered and change direction. Both cases are possible, whether energy is lost or not. Thomson scattering and Rayleigh scattering are the processes of photon interaction with matter without energy transfer. The principal mechanisms of the energy deposition of photons in matter are photoelectric absorption, Compton scattering, pair creation, and photonuclear reaction.

7.1.1 Thomson Scattering

In Thomson scattering, a free electron oscillates in response to the electric vector of an electromagnetic wave. This type of scattering is important at low energies. The oscillating electron emits a radiation with the same frequency as the incident wave. No energy loss occurs in this process. Only the angle of deflection is changed in the collision. Quantum mechanics proves Thomson scattering is an extremity of Compton scattering when incident photon energy is brought to zero.

7.1.2 Photoelectric Effect

The photoelectric effect is a phenomenon in which electrons are emitted from matter after the absorption of energy from electromagnetic radiation such as x-rays or visible light. The emitted electrons can be referred to as photoelectrons. The energy of the emitted electrons does not depend on intensities of the incident radiation, but relates to the wavelength of radiation. As the wavelength is shorter, the electrons with larger energy are emitted. To elucidate these experimental results, Einstein proposed that the incident radiation is a quantum (photon) having the energy of $E = h\nu$, and he assumed that the photoelectron is produced when an electron in matter completely absorbs a photon. The incident photon therefore disappears. The kinetic energy of a photoelectron, T, is represented by

$$T = h\nu - B \tag{7.1}$$

where B is the binding energy of an electron orbit.

EXAMPLE 7.1

What is the energy of a photoelectron emitted from lead by interaction with photons of 100 keV energy?

SOLUTION 7.1

$$B_{\mathrm{Pb}} = 88.0 \text{ keV}, \, T_e = (100 - 88) \text{ keV} = 12 \text{ keV}$$

7.1.3 Compton Scattering

Compton scattering is the decrease in energy of an x-ray or γ-ray photon when it interacts with matter. Because of the change in photon energy, it is an inelastic scattering process. The effect is important because it demonstrates that light cannot be explained purely as a wave phenomenon. Thomson scattering, the classical theory of an electromagnetic wave scattered by charged particles, cannot explain low-intensity shift in wavelength. Light must behave as if it consists of particles in order to explain the Compton scattering. Compton's experiment verified that light can behave as a stream of particle-like quanta whose energy is proportional to the frequency.

Figure 7.1 shows the scheme for Compton scattering. The quantum model of Compton scattering is that a photon with the energy $h\nu$ and the

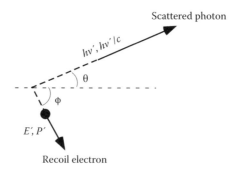

FIGURE 7.1
Definition of angles after collision.

momentum $h\nu/c$ directs at a free electron at rest. After collision, the photon is scattered through an angle θ having the energy $h\nu'$ and the momentum $h\nu'/c$, and an electron is ejected at an angle ϕ with the total energy E' and the momentum P'. The relativistic expression for E' and P' for the electron rest mass m and the velocity $\beta = v/c$ is as follows:

$$E' = mc^2 / \sqrt{1 - \beta^2} , \quad P' = mc\beta / \sqrt{1 - \beta^2} \tag{7.2}$$

From the conservation laws for energy and momentum, the following equations are obtained.

$$h\nu + mc^2 = h\nu' + E'$$

$$\frac{h\nu}{c} = \frac{h\nu'}{c} \cos\theta + P' \cos\phi \tag{7.3}$$

$$\frac{h\nu'}{c} \sin\theta = P' \sin\phi$$

Using Equations (7.2) and (7.3) the energy of the scattered photon is obtained as

$$h\nu' = \frac{h\nu}{1 + \dfrac{h\nu}{mc^2}(1 - \cos\theta)} \tag{7.4}$$

The kinetic energy of the recoiled electron, T, becomes

$$T = h\nu - h\nu' = h\nu \frac{1 - \cos\theta}{1 - \cos\theta + mc^2/h\nu} \tag{7.5}$$

If $\theta = 180°$ then the electron has the maximum energy:

$$T_{max} = \frac{2h\nu}{2 + mc^2/h\nu} \tag{7.6}$$

In the field of γ-ray spectroscopy, T_{max} is known as the Compton edge. It is found that T_{max} approaches $h\nu$ for $h\nu \gg mc^2$. The relationship between θ and ϕ is given by

$$\cot\frac{\theta}{2} = \left(1 + \frac{h\nu}{mc^2}\right)\tan\phi \tag{7.7}$$

When θ increases from $0°$ to $180°$, ϕ decreases from $90°$ to $0°$. This means that the photon is scattered through all directions, while the recoil angle of the electron is limited to the forward angles $(0 \le \phi \le 90°)$.

EXAMPLE 7.2

1. What is the maximum energy of the Compton electron for the incident photon of energy 1 MeV?
2. Calculate the wave length of the scattered photon through the angle of 45°.
3. At which angle is the Compton electron of energy 250 keV detected?

SOLUTION 7.2

1. $T_{max} = \dfrac{2 \cdot 1 \text{ MeV}}{2 + 0.511} = 0.796 \text{ MeV}$

2. $hv' = \dfrac{1 \text{ MeV}}{1 + \dfrac{1 \text{ MeV}}{0.511 \text{ MeV}}(1 - \cos 45°)} = 0.636 \text{ MeV}$

   The wavelength of the scattered photon is

   $\lambda = \dfrac{hc}{E} = \dfrac{1240 \text{ eV nm}}{0.636 \text{ MeV}} = 1.95 \cdot 10^{-3} \text{ nm}$

3. $\cos\theta = 1 - \dfrac{T \dfrac{mc^2}{hv}}{hv - T} = 1 - \dfrac{0.75 \cdot 0.511}{0.25} = -0.533 \Rightarrow \theta = 122.21°$

   $\Rightarrow \tan\phi = \dfrac{\cot\dfrac{122.21°}{2}}{(1 + 1/0.511)} = 0.187 \Rightarrow \phi = 10.57°$

7.1.4 Pair Creation

When photon energy is at least twice the electron rest mass ($hv \geq 2mc^2$), the photon is converted to a pair of electron-positron in the vicinity of a nucleus. Pair creation occurs in the field of atomic electron for the photon energy greater than $4mc^2$; however, its probability is very low. This process is called triplet pair creation because another recoiled electron is produced. In photon-nucleus pair creation, the recoiled energy of a heavy nucleus is negligible. Therefore, the photon energy hv is converted to

$$hv = 2mc^2 + T_+ + T_- \tag{7.8}$$

where T_+ and T_- represent the kinetic energies of the positron and electron, respectively. The energy distribution of the electron and positron continuously varies in the range between 0 and $hv - 2mc^2$. Those energy spectra are almost the same. The threshold energy is 1.022 MeV.

The inverse process occurs such that photons are produced by annihilation of an electron-positron pair. The positron slows down and attracts an electron. Then, a positronium that resembles a hydrogen atom is formed. This survives ~10^{-10} s before pair annihilation. The annihilation generates two photons with energy of 511 keV. These photons are emitted in opposite directions. The probability of in-flight annihilation is less than 10%.

7.1.5 Photonuclear Reaction

Photonuclear reaction is the process in which nucleons are ejected from nucleus-absorbing photons. For example, a neutron is emitted when a γ-ray is captured by a nucleus of $^{206}_{82}$Pb: the notation of this reaction is $^{206}_{82}$Pb(γ, n)$^{205}_{82}$Pb. For this to occur, photons should have sufficient energy greater than the binding energy of nucleons. Those are usually above several MeV. The kinetic energy of the emitted neutron is equal to the photon energy less the binding energy. The probability of a photonuclear reaction is much smaller in comparison with a photoelectric effect, Compton scattering, and pair creation. However, this reaction is important from the viewpoint of radiation protection because it produces neutrons. In addition, the residual nucleus after the reaction frequently becomes radioactive. From these reasons this reaction plays a significant role in the vicinity of high-energy electron accelerators. Table 7.1 shows the Q values of (γ, n) reactions that produce neutrons. The threshold energy for (γ, p) reactions is greater than that for (γ, n) reactions because higher energy is needed for protons to exceed the repulsive force of the Coulomb barrier. Reactions (γ, 2n), (γ, np), and (γ, α) occur other than (γ, *n*) and (γ, *p*).

TABLE 7.1

Q Values for (γ, n) Reactions

Reaction	Q Value (MeV)	Decay Mode of Product Nucleus
^{12}C(γ, n)^{11}C	−19.0	β^+
^{14}N(γ, n)^{13}N	−10.7	β^+
^{16}O(γ, n)^{15}O	−16.3	β^+
^{23}Na(γ, n)^{22}Na	−12.1	β^+
^{27}Al(γ, n)^{26}Al	−14.0	β^+
^{40}Ca(γ, n)^{39}Ca	−15.9	β^+
^{56}Fe(γ, n)^{55}Fe	−11.2	EC
^{63}Cu(γ, n)^{62}Cu	−10.9	β^+
^{65}Cu(γ, n)^{64}Cu	−10.2	EC(100), β^-(38), β^+(19)
^{206}Pb(γ, n)^{205}Pb	−8.25	EC
^{207}Pb(γ, n)^{206}Pb	−6.85	Stable
^{208}Pb(γ, n)^{207}Pb	−8.1	Stable

7.2 Attenuation Coefficients

The penetration of photons in matter is statistically dominated by the inter-action probability per unit length. This probability μ is called the linear attenuation coefficient or the macroscopic cross section. The dimension is the inverse of length (m^{-1}). The coefficient μ depends on both photon energy and matter. The mass attenuation coefficient is represented by μ/ρ (m^2 kg^{-1}), in which ρ is the density of matter. This denotes the interaction probability per thickness in kg.m^{-2} units. Monochromatic photons attenuate exponen-tially in a uniform target. It is assumed that the number N_0 of monochro-matic photons with a narrow beam vertically enter a slab. Some photons are scattered and other photons are absorbed in the absorber. The number of photons reached at the depth x without any interactions is assumed $N(x)$. The number of interactions dN in the small interval dx is written by

$$dN = -\mu N dx \tag{7.9}$$

in which μ is the linear attenuation coefficient. Solving this equation,

$$N(x) = N_0 e^{-\mu x} \tag{7.10}$$

is obtained. The ratio $N(x)/N_0$ is the probability that the vertically incident photons traverse a slab with the thickness x without interaction. The linear attenuation coefficient can be measured using such a setup as shown in Figure 7.2. A small size detector with the diameter d is located at the dis-tance R from the absorber ($R \gg d$). This condition is called a narrow beam or a good geometry. Only photons passed through without interaction are detected. Using Equation (7.10), the coefficient μ is obtained from measure-ments of $N(x)/N_0$ as a function of x. Figure 7.3 shows the linear attenuation

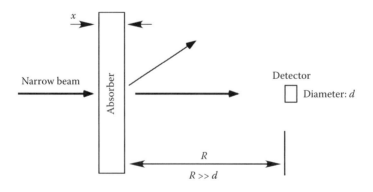

FIGURE 7.2
Measurements of attenuation coefficients using narrow beam.

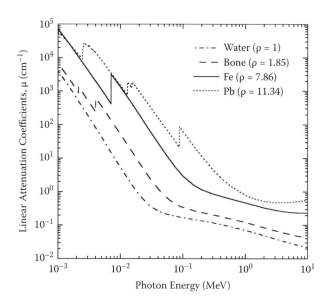

FIGURE 7.3
Linear attenuation coefficients for various matter.

coefficients for various matters for photons with energies between 1 keV and 10 MeV.

The photon linear attenuation coefficient is the sum of contributions from various processes removing photons from the narrow beam.

$$\mu = \tau + \sigma + \kappa \tag{7.11}$$

where τ, σ, and κ denote photoelectric effect, Compton effect, and pair creation, respectively. Here, Thomson scattering and photonuclear reaction are omitted. The mass attenuation coefficients for a matter with the density of ρ are τ/ρ, σ/ρ, and κ/ρ. Photoelectric effect is dominant at the lower-energy region. High Z material gives greater attenuation and absorption; however, these rapidly decrease as energies increase. Compton scattering becomes dominant above a few hundred keV. Pair creation increases above 1.022 MeV.

The linear attenuation coefficient is from photon cross sections in barn/atom units. μ is the product of the atomic density N_A by the total atomic cross section σ_A.

$$\mu = N_A \sigma_A \tag{7.12}$$

The number of atoms N_A per 1 cm^3 is given by $\rho N_0/A$ for the gram atomic mass A and Avogadro number N_0.

$$\frac{\mu}{\rho} = \frac{N_0 \sigma_A}{A} \tag{7.13}$$

This equation gives a relationship between mass attenuation coefficients and atomic photon cross sections. For compound and mixture materials, contributions of each constituent element are summed:

$$\frac{\mu}{\rho} = \frac{N_0}{A} \sum_j f^j \sigma_A^j \qquad (7.14)$$

where A is the gram molecular mass, f^j the number of the jth atom in a molecule, and σ_A^j the jth atomic cross section.

EXAMPLE 7.3

The total cross section of C for 1 MeV photons is 1.267 b. The atomic mass of C is 12.011 and the Avogadro number is 6.022×10^{23}. Calculate the mass attenuation coefficient.

SOLUTION 7.3

$$\frac{\mu}{\rho} = \frac{6.022 \times 10^{23}}{12.011} \times 1.267 \times 10^{-24} = 6.35 \times 10^{-2} \ \text{cm}^2\text{g}^{-1}$$

EXAMPLE 7.4

Total cross sections of H and O for 1 MeV photons are 0.2112 b and 1.69 b, respectively. The atomic masses of H and O are 1 and 16, respectively. Calculate the mass attenuation coefficient for water.

SOLUTION 7.4

$$\frac{\mu}{\rho} = \frac{6.022 \times 10^{23}}{18} \times (2 \times 0.2112 + 1 \times 1.69) \times 10^{-24} = 7.06 \times 10^{-2} \ \text{cm}^2\text{g}^{-1}$$

7.3 Half-Value Layer of X-Rays

The energy spectrum of x-rays used for diagnosis is not monochromatic. Since the lower-energy component is apt to be absorbed in matter, the fraction of high-energy photons rises in penetrating the matter. The attenuation for such polychromatic photons cannot be treated by counting the number of photons, while the exposure or exposure rate is measured. The attenuation is represented by the absorber (filter) thickness for which those quantities decrease by a half. This thickness is called the half-value layer (HVL). The

HVL measurement is extensively used for obtaining the properties of x-rays using narrow beams. If broad beams are used, the measured HVL is overestimated in comparison with the true value because of additional scattering from the filter. The exposure rate I for the filter thickness of x is given by

$$I = I_0 e^{-\mu x} \tag{7.15}$$

where I_0 is the rate for the case of filterless. From the definition, the HVL becomes the thickness x when $I/I_0 = 1/2$ is held.

$$\mathrm{HVL} = \frac{\ln 2}{\mu} = \frac{0.693}{\mu} \tag{7.16}$$

If the HVL for a continuous spectrum of x-rays is equal to that for monochromatic photons for a filtration material, the photon energy is called the effective energy of continuous x-rays. The voltage of the x-ray generator is represented by the peak voltage. Generally, the effective voltage (the effective energy) is less than half of the peak voltage. X-rays passing through the HVL have higher energy than before penetration; that is, they become harder. The thickness of filter necessary for degrading the exposure rate after penetration by half again is called the second half-value layer. The second HVL is thicker than the first HVL.

EXAMPLE 7.5

The mass attenuation coefficient of Al ($\rho = 2.7$ g cm^{-3}) for 60 keV photons is 0.2778 cm^2 g^{-1}. Calculate the half-value layer in mm units of Al for 60 keV photons.

SOLUTION 7.5

$$\mu = \frac{\mu}{\rho} \cdot \rho = 0.2778 \times 2.7 = 0.75 \text{ cm}^{-1} = 0.075 \text{ mm}^{-1}$$

$$\therefore \quad \mathrm{HVL} = \frac{0.693}{0.075} = 9.24 \text{ mm}$$

EXAMPLE 7.6

The mass absorption coefficient of 2.5 MeV x-rays in lead is 0.0042 m^2 kg^{-1}. Find the thickness of tissue of density 11,300 kg m^{-3} that will reduce the intensity of the radiation by a factor 10.

SOLUTION 7.6

$$I = I_0 e^{-\mu x}, \; x = (\mu_m \, \rho)^{-1} \cdot \ln I_0/I = 0.0485 \text{ m}$$

7.4 Mass Energy Absorption Coefficients

The mechanism of energy loss of photons is considered in this section. In the photoelectric effect, a secondary electron with the initial kinetic energy $T = h\nu - B$ is produced absorbing a photon energy $h\nu$. After emission of the photoelectron, the inner shell is occupied and a characteristic x-ray or an Auger electron is emitted. The fraction of energy transferred to a photoelectron or Auger electron is $1 - \delta/h\nu$, in which δ is the average energy of fluorescent x-rays. If the mass attenuation coefficient of the photoelectric effect is τ/ρ, then the mass energy transfer coefficient becomes

$$\frac{\tau_{tr}}{\rho} = \frac{\tau}{\rho}\left(1 - \frac{\delta}{h\nu}\right) \tag{7.17}$$

Because the photoelectron and Auger electron generate bremsstrahlung photons to some extent, the energy transfer coefficient does not necessarily represent energy absorption. The mass energy transfer coefficient in Compton scattering for a monochromatic photon is given by

$$\frac{\sigma_{tr}}{\rho} = \frac{\sigma}{\rho}\frac{T_{avg}}{h\nu} \tag{7.18}$$

in which $T_{avg}/h\nu$ means the average fraction converted into the initial kinetic energy of Compton electrons. Bremsstrahlungs due to Compton electrons are not taken into account here. On pair creation, the sum of the initial kinetic energy of a pair is $h\nu - 2mc^2$. The mass energy transfer coefficient is therefore given by

$$\frac{\kappa_{tr}}{\rho} = \frac{\kappa}{\rho}\left(1 - \frac{2mc^2}{h\nu}\right) \tag{7.19}$$

The total mass energy transfer coefficient becomes

$$\frac{\mu_{tr}}{\rho} = \frac{\tau}{\rho}\left(1 - \frac{\delta}{h\nu}\right) + \frac{\sigma}{\rho}\left(\frac{T_{avg}}{h\nu}\right) + \frac{\kappa}{\rho}\left(1 - \frac{2mc^2}{h\nu}\right) \tag{7.20}$$

This coefficient determines the initial kinetic energy of all electrons generated directly and indirectly. The absorbed energy at the vicinity of the interaction point is equal to the transferred energy except bremsstrahlung. Putting the average fraction of energy emitted by bremsstrahlung g, the mass energy absorption coefficient is defined by

$$\frac{\mu_{en}}{\rho} = \frac{\mu_{tr}}{\rho}(1 - g) \tag{7.21}$$

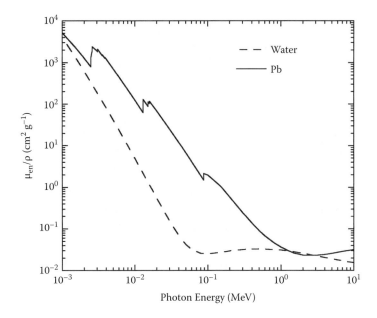

FIGURE 7.4
Comparison of mass energy absorption coefficients between water and Pb.

Figure 7.4 shows the mass energy absorption coefficients for various matter in the energy range 1 keV–10 MeV. Table 7.2 shows the differences between the coefficients for water and lead. Bremsstrahlungs for water is insignificant below 10 MeV. For lead, the difference between mass energy transfer coefficient and mass energy absorption coefficient can be explained by bremsstrahlung emission.

TABLE 7.2

Mass Attenuation Coefficients, Mass Energy Transfer Coefficients, and Mass Energy Absorption Coefficients of Photons for Water and Pb (cm^2g^{-1} units)

Photon Energy (MeV)	Water			Pb		
	μ/ρ	μ_{tr}/ρ	μ_{en}/ρ	μ/ρ	μ_{tr}/ρ	μ_{en}/ρ
0.01	5.33	4.95	4.95	131.0	126.0	126.0
0.10	0.171	0.0255	0.0255	5.55	2.16	2.16
1.0	0.0708	0.0311	0.0310	0.0710	0.0389	0.0379
10.0	0.0222	0.0163	0.0157	0.0497	0.0418	0.0325
100.0	0.0173	0.0167	0.0122	0.0931	0.0918	0.0323

The absorbed dose D in a medium is represented by the product of the energy fluence Ψ and the energy absorption coefficient μ_{en}/ρ under the condition of electron equilibrium, such as

$$D = \Psi \frac{\mu_{en}}{\rho} \qquad (7.22)$$

A similar equation for the kerma K holds using the mass energy transfer coefficient μ_{tr}/ρ,

$$K = \Psi \frac{\mu_{tr}}{\rho} \qquad (7.23)$$

This relationship does not need the condition of electron equilibrium.

EXAMPLE 7.7

The mass attenuation coefficient of W for a photoelectric effect for 100 keV photons is 4.15 cm^{2}g^{-1}. The average energy of K-x-rays is 60.6 keV. Calculate the mass energy transfer coefficient for the photoelectric effect for 100 keV photons.

SOLUTION 7.7

$$\frac{\tau_{tr}}{\rho} = 4.15 \times \left(1 - \frac{60.6}{100}\right) = 1.64 \text{ cm}^2\text{g}^{-1}$$

EXAMPLE 7.8

The mass attenuation coefficient of water for Compton scattering for 0.5 MeV photons is 0.0966 cm^2 g^{-1}. The average energy of scattered photons is 0.329 MeV. Calculate the mass energy transfer coefficient for Compton scattering for 0.5 MeV photons.

SOLUTION 7.8

The average energy of recoil electrons is 0.5 − 0.329 = 0.171 MeV.

$$\frac{\sigma_{tr}}{\rho} = 0.0966 \times \frac{0.171}{0.5} = 0.033 \text{ cm}^2\text{g}^{-1}$$

EXAMPLE 7.9

The mass energy transfer and mass energy absorption coefficients are 0.0271 and 0.0270 cm^2 g^{-1}, respectively. Calculate the average fraction of energy emitted by bremsstrahlung.

SOLUTION 7.9

$$g = 1 - \frac{\mu_{en}}{\mu_{tr}} = 1 - \frac{0.0270}{0.0271} = 0.0037$$

EXAMPLE 7.10

The mass energy absorption coefficient of Al for 80 keV photons is $0.05511 \text{ cm}^2\text{g}^{-1}$. If the fluence is $1 \times 10^6 \text{ cm}^{-2}$, what is the absorbed energy?

SOLUTION 7.10

$$D = 80 \times 10^6 \times 0.05511 = 4.41 \times 10^6 \text{ keVg}^{-1}$$

Expressed in SI units,

$$D = \frac{4.41 \times 10^6 \text{ keV}}{g} \times 1.6 \times 10^{-16} \frac{J}{keV} \times 10^3 \frac{g}{kg} = 7.06 \times 10^{-7} \text{ Gy}$$

7.5 Summary

1. Photon interactions with matter are in the form of Thomson scattering, photoelectric effect, Compton scattering, electron pair creation, and photonuclear reaction.

2. The major process of energy deposition of photons in matter is photoelectric absorption in the low-energy region, Compton scattering in the medium-energy region, and electron pair creation in the high-energy region.

3. A photon travels a distance called the free path before interaction with an atom. The mean free path is determined by the attenuation coefficient.

4. The thickness of absorber that reduces a dose by a half is called the half-value layer (HVL). The HVL is used for representing radiation quality of the continuous x-rays.

5. Under the condition that electron equilibrium holds, the absorbed dose at a point in the medium is equal to the product of energy fluence by mass energy absorption coefficient.

QUESTIONS

1. a. Write an expression for the energy relation between the incident and the scattered photon.
   b. Write a corresponding relation for the electron.
   c. Derive an expression relating the energy of an incident and a scattered photon in a Compton process.
2. Calculate the specific γ-ray constant for ^{60}Co in units R per h per Ci at a distance of 1 m.
3. In a Compton interaction the scattered photon gets three times more energy than the ejected electron. Calculate the minimum incident photon energy.
4. What is the energy of the back scattered photon from Compton interaction of ^{137}Cs γ (662 keV)?
5. Calculate ^{137}Cs γ-ray attenuation by 1 cm lead and aluminum.
6. Neutron contamination from accelerators in radiotherapy could cause secondary cancer. The photoneutron could be produced in the head of the accelerators when a photon hits the target, or collimators. The target and collimators are made of heavy atomic number materials like tungsten and lead. What is the minimum energy of the photon in order to avoid contamination of the neutron?
7. Derive the minimum photon energy that can induce triplet production.
8. Calculate the number of interactions that occur in lead 15 mm thick impinged perpendicularly by a narrow photon beam with 10^{15} photons of energy 10 MeV. The density of lead is 11.3 g/cm^2.

For Further Reading

ICRU. 1992. *Photon, Electron, Proton and Neutron Interaction Data for Body Tissues.* ICRU Report 46.
Storm E, Israel HI. 1970. Photon cross sections from 1 keV to 100 MeV for elements $Z = 1$ to $Z = 100$. *Nuclear Data Tables* A7: 565–681.

8

Interaction of Electrons with Matter

8.1 Energy Loss of Charged Particles

When charged particles penetrate matter, their energies are lost by ionization and excitation of atoms or molecules of the medium. Moving charged particles act on atomic electrons through electromagnetic force and transfer energy to these electrons. The transferred energy may be sufficient for ionizing the atom by ejecting an orbital electron or may produce an excited state not ionized. Heavy charged particles lose a small part of their energies by a single collision. The deflection of the path when collision occurs can be ignored. Heavy charged particles travel almost in a straight line, continuously losing a small amount of energies by collision with the atomic electrons. Ionized or excited atoms are left on the track of penetration. Sometimes, the path is bent by the Rutherford scattering with the nucleus. Electrons and positrons lose energies almost continuously when being slowed down in matter. They lose a major part of energies by a single collision with the orbital electron. The path is largely bent because the scattering angle determined from the kinematics is usually large. Electrons are largely scattered by elastic scattering with the nucleus. Electrons and positrons do not advance in a straight line. In addition, electrons emit bremsstrahlung photons when the path is sharply bent. The contribution of bremsstrahlung to the stopping power becomes important at high-energy regions. For example, the radiative stopping power for water for 100 MeV electrons contributes almost half to the total stopping power.

The mean energy loss of charged particles per length in medium is important in the fields of radiation physics and radiation dosimetry. This quantity, noted by $-dE/dx$, is called the stopping power of the medium for the particle—or from the viewpoint of the particle, it is frequently represented by the linear energy transfer (LET). The unit of LET is usually keV μm^{-1}. The stopping power and LET are closely related to the dose given by the recoiled charged particles produced by the uncharged particles, such as photons and neutrons. In addition, those are related to the biological effect of various radiations. The stopping power is defined by the product of the mean energy loss per collision, Q_{av}, and the collision probability per unit length, μ, in which μ is the macroscopic cross section whose dimension is the inverse of the length. The mean energy loss Q_{av} is then given by

$$Q_{av} = \int_{Q_{min}}^{Q_{max}} QW(Q)dQ \tag{8.1}$$

where $W(Q)$ is the energy loss spectrum for a collision. The minimum energy loss, Q_{min}, in the collision between charged particle and electron seems to be ~0. The maximum energy loss, Q_{max}, can be roughly estimated from the kinematical relationship. If a charged particle having the mass M and the speed V collides with an electron having the mass m in rest, the energy transfer reaches maximum for the head-on collision. Solving the conservation laws of energy and momentum, Q_{max} is obtained as

$$Q_{max} = \frac{4mME}{(M+m)^2} \tag{8.2}$$

in which $E = MV^2/2$ is the initial kinetic energy of the charged particle. If the incident particle is an electron or positron, Q_{max} is E. It is experimentally confirmed that the energy transfer continuously distributes in the range $Q_{min} < Q < Q_{max}$ and Q_{av} is ~20 eV.

The linear stopping power is defined by

$$-\frac{dE}{dx} = \mu Q_{av} = \mu \int_{Q_{min}}^{Q_{max}} QW(Q)dQ \tag{8.3}$$

The unit of the linear stopping power is generally MeV cm^{-1}. The quantity divided by the material density, $-dE/(\rho dx)$, is called the mass stopping power, whose unit is MeV cm^2 g^{-1}. The values of the stopping power vary according to particle, energy, and medium.

EXAMPLE 8.1

Calculate the maximum energy that a 5 MeV proton can lose in a single electronic collision.

SOLUTION 8.1

Neglecting m compared with M, the maximum energy transfer can be approximated by

$$Q_{max} = \frac{4mE}{M} = \frac{4 \times 1 \times 5}{1836} = 1.09 \times 10^{-2} \quad \text{MeV} = 10.9 \text{ keV}$$

8.2 Collision Stopping Power

We call both an electron and a positron an electron because their stopping powers and ranges are almost identical. The stopping powers of electrons are different from those of charged particles for two reasons. The first is that an electron loses a great deal of energy by a single collision. The second is that it is impossible to identify either an incident or hit electron. On the energy loss, the higher-energy electron after collision is regarded as the incident electron. Therefore, the maximum energy transfer becomes $T/2$ for the negatively charged electron with the kinetic energy T.

The stopping power formula derived from the quantum mechanics is summarized in ICRU Report 37. The mass collision stopping power is given by

$$\frac{S_{col}^{\mp}}{\rho} = \frac{2\pi r_e^2 mc^2}{\beta^2} \frac{ZN_A}{A} \left[\ln\left(\frac{T}{I}\right)^2 + \ln\left(1+\frac{\tau}{2}\right) + F^{\mp}(\tau) - \delta \right] \tag{8.4}$$

$$F^-(\tau) = (1-\beta^2)\left[1 + \frac{\tau^2}{8} - (2\tau+1)\ln 2 \right] \qquad \text{for electrons} \tag{8.5}$$

$$F^+(\tau) = 2\ln 2 - \frac{\beta^2}{12}\left[23 + \frac{14}{\tau+2} + \frac{10}{(\tau+2)^2} + \frac{4}{(\tau+2)^3} \right] \qquad \text{for positrons} \tag{8.6}$$

in which T = kinetic energy of electron, mc^2 = rest energy of electron, $\tau = T/mc^2$, $\beta = V/c$ = ratio of electron speed and light speed, $r_e = e^2/mc^2$ = classical electron radius, Z = atomic number of target atom, A = atomic mass of target atom, N_A = Avogadro number, and I = mean excitation energy of medium.

In liquid and solid mediums, the density effect appears, which decreases the energy loss because of the polarization of the medium affected by the electric field of the incident charged particle. The last term in Equation (8.4), δ, represents the density effect. S_{col} omitted ρ is called the linear collision stopping power. The subscript col is added to discriminate the radiative stopping power S_{rad}/ρ, described in the next section. The mean excitation energy I for various elements is obtained from the following approximations:

$$I \approx 19.0 \quad \text{eV}, \quad Z = 1$$

$$11.2 + 11.7Z \quad \text{eV}, \quad 2 \leq Z \leq 13 \tag{8.7}$$

$$52.8 + 8.71Z \quad \text{eV}, \quad Z > 13$$

The mass collision stopping power for compounds and mixtures can be approximated by the linear combination of the constituent atoms.

$$\frac{S_{col}}{\rho} = \sum_j w_j \left(\frac{S_{col}}{\rho} \right)_j \tag{8.8}$$

where w_j is the weight ratio of the jth atom. The corresponding mean excitation energy is given by

$$\ln I = \frac{\sum_j w_j (Z_j/A_j) \ln I_j}{\sum_j w_j (Z_j/A_j)} \tag{8.9}$$

Figure 8.1 shows the linear collision stopping powers of various matters for electrons. Table 8.1 shows the various data regarding the stopping power of water for electrons. Although electrons lose at most half of the energy in a collision, positrons have the possibility of losing all energy. Therefore, the collision stopping powers for positrons slightly differ from those of electrons. The collision stopping powers for positrons are ~98% those of electrons in the energy range greater than 500 keV. In opposition, those are larger than electron data by ~5% at 100 keV and 10 ~ 20% at 10 keV.

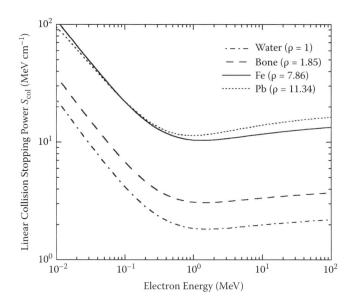

FIGURE 8.1
Linear collision stopping powers of various media for electrons.

TABLE 8.1

Mass Collision Stopping Powers, Mass Radiative Stopping Powers, Radiation Yields, and Ranges of Water for Electrons

Kinetic Energy (MeV)	β^2	$\dfrac{S_{col}}{\rho}$	$\dfrac{S_{rad}}{\rho}$ MeV cm² g⁻¹	$\dfrac{S_{tot}}{\rho}$	Radiation Yield	Range (g cm⁻²)
0.001	0.0039	126.		126.		5×10^{-6}
0.002	0.00778	77.5		77.5		2×10^{-5}
0.005	0.0193	42.6		42.6		8×10^{-5}
0.010	0.0380	23.2		23.2	0.0001	0.0002
0.025	0.0911	11.4		11.4	0.0002	0.0012
0.050	0.170	6.75		6.75	0.0004	0.0042
0.075	0.239	5.08		5.08	0.0006	0.0086
0.1	0.301	4.20		4.20	0.0007	0.0140
0.2	0.483	2.84	0.006	2.85	0.0012	0.0440
0.5	0.745	2.06	0.010	2.07	0.0026	0.174
0.7	0.822	1.94	0.013	1.95	0.0036	0.275
1	0.886	1.87	0.017	1.89	0.0049	0.430
4	0.987	1.91	0.065	1.98	0.0168	2.00
7	0.991	1.93	0.084	2.02	0.0208	2.50
10	0.998	2.00	0.183	2.18	0.0416	4.88
100	0.999	2.20	2.40	4.60	0.317	32.5

EXAMPLE 8.2

Calculate the collisional stopping power of water ($I = 75$ eV) for 1 MeV electrons assuming $\delta = 0$.

SOLUTION 8.2

Using $\tau = \frac{1}{0.511} = 1.96$ and $\beta^2 = 1 - (\frac{0.511}{1+0.511})^2 = 0.886$,

$$F^- = -0.22$$

$$r_e^2 = 7.94 \times 10^{-26}, \ N_A = 6.022 \times 10^{23}, \ Z = 10, \ A = 18.015,$$

$$\ln\left(\frac{10^6}{75}\right)^2 = 18.996, \ \ln\left(1 + \frac{1.96}{2}\right) = 0.683$$

$$\frac{S^-}{\rho} = \frac{2\pi \times 7.941 \times 10^{-26} \times 0.511}{0.886} \cdot \frac{10 \times 6.022 \times 10^{23}}{18.015} [18.996 + 0.683 - 0.22]$$

$$= 1.87 \text{ MeV cm}^2 \text{ g}^{-1}$$

EXAMPLE 8.3

Calculate the collisional stopping power of water ($I = 75$ eV) for 1 MeV positrons assuming $\delta = 0$.

SOLUTION 8.3

Using $F^+ = -0.625$,

$$\frac{S^+}{\rho} = \frac{2\pi \times 7.941 \times 10^{-26} \times 0.511}{0.886} \cdot \frac{10 \times 6.022 \times 10^{23}}{18.015}[18.996 + 0.683 - 0.625]$$

$$= 1.83 \text{ MeV cm}^2 \text{ g}^{-1}$$

EXAMPLE 8.4

Calculate the mean excitation energy of SiO_2.

SOLUTION 8.4

$$I_{Si} = 52.8 + 8.71 \times 14 = 174.7, \qquad I_O = 11.2 + 11.7 \times 8 = 104.8$$

$$w_{Si} = \frac{28}{28 + 16 \times 2} = 0.467, \qquad w_O = \frac{16 \times 2}{28 + 16 \times 2} = 0.533$$

$$\ln I = \frac{0.467 \times 0.5 \times \ln 174.7 + 0.533 \times 0.5 \times \ln 104.8}{0.467 \times 0.5 + 0.533 \times 0.5} = 4.891$$

$$\therefore \qquad I_{SiO_2} = 133 \text{ eV}$$

8.3 Radiative Stopping Power

Bremsstrahlung emitted from the collision of heavy charged particles with an atom can be ignored because particles are not accelerated. On the other hand, electrons receive strong acceleration and emit bremsstrahlung. The definition of mass radiative stopping power is given by

$$\frac{S_{rad}}{\rho} = \frac{N_A}{A} \int_0^T k \frac{d\sigma}{dk} dk \qquad (8.10)$$

where $d\sigma/dk$ is the differential cross section that emits a photon energy k due to the electron interaction with the nuclear Coulomb field. It is possible to evaluate the radiative stopping powers if the cross section is known. However, it is difficult to represent $d\sigma/dk$ using a unique analytical formula.

Instead, numerical calculations are done dividing the regions into low, intermediate, and high energies.

We see that the bremsstrahlung efficiency is in proportion to Z^2 and linearly increases depending on electron energy. The collision stopping powers for high-energy electrons show the logarithmic increase. Bremsstrahlung becomes a dominant mechanism of energy loss at high energies. The next approximation provides the ratio between radiative loss and collision loss for the electron with the total energy E (MeV).

$$\frac{S_{rad}}{S_{col}} \cong \frac{ZE}{800} \tag{8.11}$$

For lead ($Z = 82$), the energy for which both stopping powers are identical is $E = 9.8$ MeV; $T = E - mc^2 = 9.3$ MeV is obtained. The total mass stopping power is the sum of two losses such as

$$\frac{S_{tot}}{\rho} = \frac{S_{col}}{\rho} + \frac{S_{rad}}{\rho} \tag{8.12}$$

Figure 8.2 shows the radiative and total stopping powers of water and Pb. The radiation yield Y represents a fraction of energy dissipated for bremsstrahlung emission until electrons completely stop. For 100 MeV electrons in

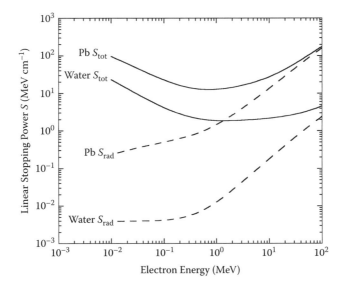

FIGURE 8.2
Linear radiative stopping powers and linear total stopping powers of water and Pb for electrons.

water, Y becomes 31.7%. If electrons of the initial kinetic energy T (MeV) stop in the absorber of the atomic number Z, Y is approximated by

$$Y \approx \frac{6 \times 10^{-4} ZT}{1 + 6 \times 10^{-4} ZT} \tag{8.13}$$

Bremsstrahlung for positrons is almost the same as for electrons at high energies, but becomes smaller at low energies.

EXAMPLE 8.5

Calculate the energy for which both stopping powers are identical for water.

SOLUTION 8.5

Using $Z = 10$ for water, $E = 80$ MeV and $T = 79.5$ MeV.

EXAMPLE 8.6

Calculate the radiation yield for 10 MeV electrons in Pb.

SOLUTION 8.6

$$Y = \frac{6 \times 10^{-4} \times 82 \times 10}{1 + 6 \times 10^{-4} \times 82 \times 10} \times 100 = 33\%$$

8.4 Ranges

The distance that charged particles advance until stop is called the range. It is assumed that charged particles continuously slow down until the initial energy is completely lost. This is called the continuous slowing down approximation (CSDA) range. This approximation is not realistic for electrons because they lose a considerable fraction of the energy by a single collision. In addition, the electron path is tortuous, unlike heavy particles. However, the term *electron range* means the CSDA range defined by the next equation. That defines the mean path length extended from the start to the end of penetration. This is distinguished from the distance covered along the path of the electron through the absorber. The CSDA range $R(T)$ is written by

$$R(T) = \int_0^T [S_{col}(T) + S_{rad}(T)]^{-1} \, dT \tag{8.14}$$

in which the unit is cm. If multiplying ρ, then the unit changes to g cm^{-2}. Figure 8.3 shows the CSDA range in cm units for electrons in various matters.

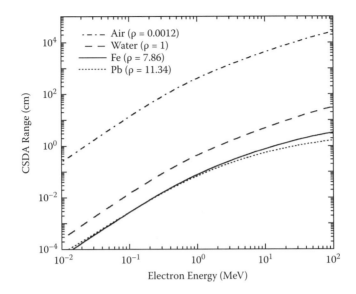

FIGURE 8.3
CSDA ranges of electrons in various matter.

The collision stopping powers of the material with large Z are smaller than those of water. Therefore, the range below 20 MeV for lead is larger in comparison with that for water. At the high-energy region, the range for lead shortens because of the increase of bremsstrahlung loss.

The empirical formula for the electron range for small Z materials is available.

$$R = 0.412T^{1.27-0.0954\ln T} \quad \text{for} \quad 0.01 \le T \le 2.5 \text{ MeV}$$

$$= 0.530T - 0.106 \quad \text{for} \quad T > 2.5 \text{ MeV} \tag{8.15}$$

in which the range is in g cm^{-2} unit and the kinetic energy in MeV. In the measurement, the practical range R_p is obtained from the absorption curve for the material. For high-energy electron beams used for radiation therapy, the practical range in water is measured under the condition of the fields 10×10 cm^2, or 10 cm diameter. The relationship is obtained as

$$R_p = 0.52T - 0.30 \tag{8.16}$$

EXAMPLE 8.7

Calculate the range of a 2 MeV electron in water.

SOLUTION 8.7

$$R = 0.412 \times 2^{1.27-0.0954\ln 2} = 0.95 \text{ cm}$$

8.5 Multiple Scattering

The reasons of nonlinearity of the electron path are: (1) a large scattering angle at the inelastic scattering with an orbital electron, and (2) a large deflection by the elastic scattering with a nucleus. If the kinetic energy T of the incident electron degrades T' by an inelastic scattering, the scattering angle θ is determined from the kinematical relationship,

$$\cos\theta = \sqrt{\frac{T'(T+2mc^2)}{T(T'+2mc^2)}} \tag{8.17}$$

EXAMPLE 8.8

Calculate the scattering angle (in degrees) when the incident electron energy of 1 MeV degrades to 0.99 MeV by an inelastic scattering.

SOLUTION 8.8

$$\cos\theta = \sqrt{\frac{0.99\times(1+1.022)}{0.99+1.022}} = 0.99746 \qquad \therefore \quad \theta = 4.1°$$

On the other hand, electron cross sections scattered by a nucleus taking into account the screening effect of orbital electrons is represented by Moliere's formula. The detailed expressions will be given in Section 13.3. This formula is for a single scattering. If the material layer is thick, a large number of scattering is repeated. This is called the multiple scattering. The cause of back scattering for electrons is the multiple scattering. This theory treats the averaged angular distributions overlapping a large number of elastic scattering in the material with finite thickness. The relative scattering intensity at the θ direction is written by

$$I(\theta) \sim \sqrt{\theta\sin\theta} \times \left(f^{(0)} + \frac{f^{(1)}}{B} + \frac{f^{(2)}}{B^2} \right) \tag{8.18}$$

in which $f^{(0)}$, $f^{(1)}$, $f^{(2)}$ are functions of the reduced angle Θ. Table 8.2 shows the numerical values for them given by Bethe. The parameters Θ and B are calculated using the electron radius r_e, the Avogadro constant N_A, the density of material ρ, and the thickness of layer t. For the ith atom in compound materials or mixtures, the atomic number, the atomic mass, and the number of atoms in a mole are Z_i, A_i, and p_i, respectively.

$$\Theta = \frac{\theta}{\chi_c\sqrt{B}} \tag{8.19}$$

$$\chi_c^2 = \frac{4\pi r_e^2 N_A \rho t Z_S(1-\beta^2)}{A\beta^4} \tag{8.20}$$

TABLE 8.2

Reduced Angles and Parameters for Electron Multiple Scattering

Θ	$f^{(0)}$	$f^{(1)}$	$f^{(2)}$
0	2	0.8456	2.4929
0.2	1.9216	0.7038	2.0694
0.4	1.7214	0.3437	1.0488
0.6	1.4094	−0.0777	−0.0044
0.8	1.0546	−0.3981	−0.6068
1	0.7338	−0.5285	−0.6359
1.2	0.4738	−0.4770	−0.3086
1.4	0.2817	−0.3183	0.0525
1.6	0.1546	−0.1396	0.2423
1.8	0.0783	−0.0006	0.2386
2	0.0366	0.0782	0.1316
2.2	0.01581	0.1054	0.0196
2.4	0.00630	0.1008	−0.0467
2.6	0.00232	0.08262	−0.0649
2.8	0.00079	0.06247	−0.0546
3	0.00025	0.0455	−0.03568
3.2	7.3×10^{-5}	0.03288	−0.01923
3.4	1.9×10^{-5}	0.02402	−0.00847
3.6	4.7×10^{-6}	0.01791	−0.00264
3.8	1.1×10^{-6}	0.01366	0.00005
4	2.3×10^{-7}	0.010638	0.001074
4.5	3×10^{-9}	6.14×10^{-3}	0.001229
5	0	3.831×10^{-3}	8.326×10^{-4}
5.5	0	2.527×10^{-3}	5.368×10^{-4}
6	0	1.739×10^{-3}	3.495×10^{-4}
7	0	9.080×10^{-4}	1.584×10^{-4}
8	0	5.211×10^{-4}	7.830×10^{-5}
9	0	3.208×10^{-4}	4.170×10^{-5}
10	0	2.084×10^{-4}	2.370×10^{-5}

and

$$B - \ln B = \ln \Omega_0 \tag{8.21}$$

$$\Omega_0 = 6702.33 \frac{Z_S e^{Z_E/Z_S} \rho t}{e^{Z_X/Z_S} A \beta^2} \tag{8.22}$$

where

$$A = \sum p_i A_i$$

$$Z_S = \sum p_i Z_i (Z_i + 1)$$

$$Z_E = \sum p_i Z_i (Z_i + 1) \ln Z_i^{-2/3} \tag{8.23}$$

$$Z_X = \sum p_i Z_i (Z_i + 1) \ln \left[1 + 3.34 \left(\frac{Z_i}{137\beta} \right) \right]$$

B is obtained solving Equation (8.22) numerically. Moliere's theory holds for electrons of > 10 keV and for the thickness in which elastic scattering occurs more than 20 times. Figure 8.4 shows the angular distributions of electron multiple scattering in Al for various energies.

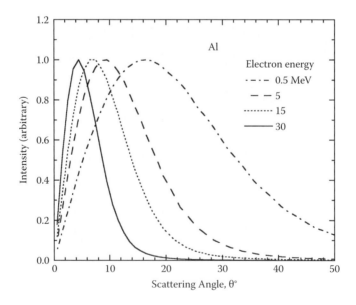

FIGURE 8.4
Angular distributions of electron multiple scattering in Al for various energies.

EXAMPLE 8.9

Calculate the angular distribution of multiple scattering, $I(\theta)$, for 1 MeV electrons in Al ($\rho = 2.69$ g cm^{-3}) setting the step size = 2% and the cutoff 10 keV for both electrons and photons.

SOLUTION 8.9

Using $\beta = 0.94108$, $A = 26.98$, $Z_S = 182$, $Z_E = -311.2$, and $Z_X = 6.078$ are obtained from Equation (8.23).

The thickness of layer, t, can be calculated by $t = \Delta E/S$, in which $\Delta E = 1,000 \times 0.02 = 20$ keV and the restricted stopping power $S = 3,097$ keV/cm. Therefore, $t = 6.458 \times 10^{-3}$ cm. From Equations (8.20) and (8.22), $\chi_c = 0.1662$ and $\Omega_0 = 155.15$ are obtained. From Equation (8.21), $B = 6.988$ is obtained. Using Equation (8.19) and Table 8.2, the angular distribution is calculated as a function of θ. A part of the results is shown in the following table.

$\theta°$	Θ	$f^{(0)}$	$f^{(1)}/B$	$f^{(2)}/B^2$	$I(\theta)$
1	0.0397	1.9844	0.1170	0.0493	0.0375
3	0.1192	1.9533	0.1089	0.0459	0.1104
5	0.1986	1.9221	0.1009	0.0424	0.1801
7	0.2781	1.8435	0.0806	0.0342	0.2389
9	0.3575	1.7639	0.0601	0.0259	0.2900
11	0.4370	1.6637	0.0380	0.0175	0.3291
13	0.5164	1.5398	0.0141	0.0089	0.3531
15	0.5959	1.4159	−0.0099	0.00036	0.3661
17	0.6753	1.2758	−0.0284	−0.0047	0.3660
19	0.7548	1.1349	−0.0466	−0.0096	0.3544
21	0.8342	0.9997	−0.0602	−0.0125	0.3360
23	0.9137	0.8723	−0.0676	−0.0128	0.3136
25	0.9931	0.7449	−0.0750	−0.0130	0.2821

8.6 Cerenkov Radiation

When a charged particle passes through an insulator at a constant speed greater than the speed of light in that medium, a visible blue glow is emitted. This phenomenon is called Cerenkov radiation, which is analogous to a shock wave of a supersonic aircraft. As the charged particle travels, it disrupts the local electromagnetic field in the medium. Atomic electrons of the medium will be displaced, and the atoms become polarized by the passing

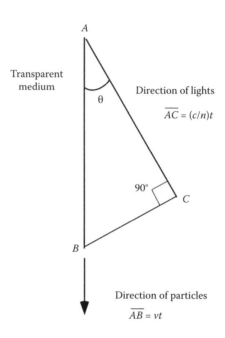

FIGURE 8.5
Principle of Cerenkov radiation.

electromagnetic field of a charged particle. Photons are emitted as an insulator's electrons restore themselves to equilibrium after the disruption has
passed. Assuming the speed of light in vacuum, c, the velocity of a charged
particle, $v = \beta c$, and the refractive index of the medium, n, the speed of light
becomes c/n. If $v > c/n$, then Cerenkov radiation is emitted. In Figure 8.5 a
particle travels the direction from point A to point B. The distance is vt. The
electromagnetic wave emitted from A reaches C within the time t. Namely,
$\overline{AC}$ becomes $(c/n)t$. Therefore, the light wave is emitted with the conical pattern having the polar angle θ satisfying

$$\cos\theta = \frac{\overline{AC}}{\overline{AB}} = \frac{(c/n)t}{vt} = \frac{1}{\beta n} < 1 \qquad (8.24)$$

The threshold kinetic energy that enables emitting of Cerenkov light for
the particle with the rest mass m is derived using the condition $\beta n > 1$ and

$$\beta = \sqrt{1 - [mc^2/(T + mc^2)]^2}$$

$$T = mc^2 \left(\frac{n}{\sqrt{n^2 - 1}} - 1 \right) \qquad (8.25)$$

This is called the critical energy. For electrons in water or lucite, those are 257 and 175 keV, respectively. The contribution of Cerenkov radiation to the electron stopping power can be neglected.

8.7 Summary

1. A charged particle moving in matter affects the electromagnetic force on the atomic electron. The particle loses energy by ionizing and exciting the atom.

2. The mean energy loss of charged particles per a unit length in medium is called the stopping power of the medium for the particles.

3. The total mass stopping power for electrons is the sum of the mass collision stopping power and mass radiative stopping power.

4. The electron path is significantly bent by the collision with the atomic electron and the multiple scattering with the nucleus. Electrons do not move straight forward. However, the range is defined by the continuous slowing down approximation.

5. The threshold energy of Cerenkov radiation for electrons in water is 257 keV.

QUESTIONS

1. Calculate the I value of water from those of H and O ($I_H = 19.2$ eV and $I_O = 95$ eV) and comment on the validity of the method.

2. Estimate how many interactions occur along a 1 μm track of a 1 MeV proton in liquid water, given that the mass stopping power for a 1 MeV proton in liquid water is 260.6 MeV cm^2 g^{-1} (NIST). How large is the electronic cross section from this estimation?

3. For a 1 cGy dose deposited in a cell layer of 1 μm thick, how many electrons of energy 1 MeV need to traverse the cell layer, assuming the cell is a water-equivalent cube?

4. How is the collisional stopping power of aluminum compared to the corresponding value of lead for 50 MeV electrons?

5. Calculate the thickness of an aluminum shield for a ^{90}Sr/^{90}Y β source.

6. β-Electrons with unknown maximum energy are emitted. If at least 1 cm plastic with the density of 1.3 g cm^{-3} is required to stop the electrons, calculate the maximum energy.

7. A 550 keV electron pencil beam travels from the top angle of a plastic cone, along the axis of the cone. The cone opening angle is 40°. Calculate the refractive index of the plastic, if the Cerenkov radiation passes through the bottom of the cone parallel to the cone axis.

For Further Reading

Bethe HA. 1953. Moliere's theory of multiple scattering. *Phys. Rev.* 89: 1256–1266.
Turner JE. 1995. *Atoms, Radiation, and Radiation Protection*, 2nd ed. New York: John Wiley & Sons.

9

Interaction of Heavy Charged Particles with Matter

9.1 Collision Stopping Powers

The mass collision stopping power formula for heavy charged particles was derived by H.A. Bethe using relativistic quantum mechanics.

$$\frac{S_{col}}{\rho} = -\frac{dE}{\rho dx} = \frac{4\pi r_e^2 mc^2 z^2 Z N_A}{\beta^2 A}\left[\ln\frac{2mc^2\beta^2}{I(1-\beta^2)} - \beta^2\right] \tag{9.1}$$

where $\beta = V/c$ is the ratio of particle velocity to the velocity of light. This is represented, if the mass of particle, M, and the kinetic energy, T, are used, as

$$\beta = \sqrt{1-\left(\frac{Mc^2}{T+Mc^2}\right)^2} \tag{9.2}$$

and mc^2 = rest energy of an electron, $r_e = e^2/mc^2$ = classical electron radius, z = charge of heavy charged particle, Z = atomic number of the target atom, A = atomic mass of the target atom, N_A = Avogadro constant, and I = mean excitation energy of medium.

The reliability of Bethe's formula was confirmed in comparison with the experimental data. However, that cannot reproduce the experimental data in the low-energy region below a few hundred keV. When fast ions slow down around the Bragg peak (~0.3 MeVu^{-1}), interactions involving electron capture and loss by the moving ions become an increasingly important component of the energy loss process. The effect of charge exchanges is corrected introducing a parameter called the effective charge. ICRU Report 49 recommends the formula adding various correction terms.

$$\frac{S_{col}}{\rho} = \frac{4\pi r_e^2 mc^2 z^2 Z N_A}{\beta^2 A} L(\beta) \tag{9.3}$$

$$L(\beta) = L_0(\beta) + z L_1(\beta) + z^2 L_2(\beta) \tag{9.4}$$

L is called the stopping number. The first term is

$$L_0(\beta) = \frac{1}{2}\ln\left(\frac{2mc^2\beta^2 Q_{max}}{1-\beta^2}\right) - \beta^2 - \ln I - \frac{C}{Z} - \frac{\delta}{2} \tag{9.5}$$

where C and δ represent the shell correction and the density effect, respectively. Q_{max} is the maximum energy loss at a collision with a free electron.

$$Q_{max} = \frac{2mc^2\beta^2}{1-\beta^2} \frac{1}{1 + \frac{2m}{M\sqrt{1-\beta^2}} + \left(\frac{m}{M}\right)^2} \tag{9.6}$$

This quantity becomes $Q_{max} = 4mT/M$ at the nonrelativistic limit. The second and third terms of Equation (9.4) are called the Barkas correction and Bloch correction, respectively. Figure 9.1 shows the linear collision stopping powers of various materials for protons. Figure 9.2 shows those for α-particles. In the low-energy region the term in front of the bracket increases for $\beta \to 0$, while the logarithmic term decreases. Consequently, the maximum stopping powers called the Bragg peak appear at around 100 keV for protons and 600 keV for α-particles.

A relationship between S and z holds for a velocity V regardless of the particle type.

$$S/z^2 = \text{constant} \tag{9.7}$$

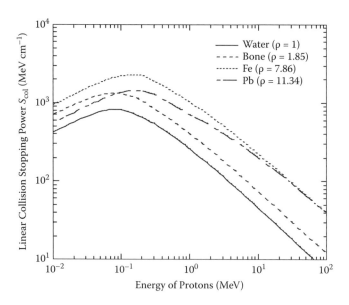

FIGURE 9.1
Linear collision stopping powers of various matters for protons.

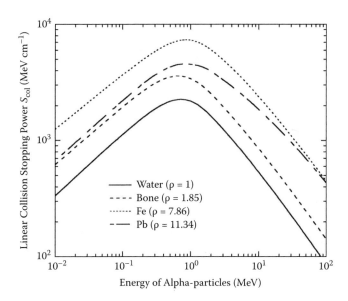

FIGURE 9.2
Linear collision stopping powers of various matters for α-particles.

If the stopping power of a particle is known, it is possible to estimate the stopping power of a different particle with the same velocity. For instance, it is assumed protons and heavy particles are moving with the same velocity *V*. The kinetic energies are given by

$$T_p = \frac{1}{2} m_p V^2, \quad T_{HI} = \frac{1}{2} m_{HI} V^2 \tag{9.8}$$

Then,

$$T_{HI} = T_p \frac{m_{HI}}{m_p} \tag{9.9}$$

is obtained. If HI = α-particles, the velocity of 4 MeV α-particles is equal to that of 1 MeV protons. From Equation (9.7), S_α is represented using S_p.

$$S_\alpha = S_p \left(\frac{z_\alpha}{z_p} \right)^2 = 4 S_p \tag{9.10}$$

Therefore, the collision stopping power of 4 MeV α-particles is equal to four times that of 1 MeV protons. This scaling law is effective for the high-energy region in which the charge exchange processes can be neglected. The collision stopping powers for particles other than α-particles are calculable from the scaling law, as shown in Table 9.1.

TABLE 9.1

Scaling Law for Mass Collision Stopping Powers for Heavy Charged Particles

		Deuteron		$^3\text{He}^{2+}$		α		$^{12}\text{C}^{6+}$	
		T	S	T	S	T	S	T	S
T_p	S_p	$T_p \times 2$	$S_p \times 1$	$T_p \times 3$	$S_p \times 4$	$T_p \times 4$	$S_p \times 4$	$T_p \times 12$	$S_p \times 36$
1	260.6	2	260.6	3	1,042	4	1,042	12	9,382
2	158.5	4	158.5	6	634.0	8	634.0	24	5,706
3	117.1	6	117.1	9	468.4	12	468.4	36	4,216
4	94.0	8	94.0	12	376.0	16	376.0	48	3,384
5	79.1	10	79.1	15	316.4	20	316.4	60	2,848
10	45.6	20	45.6	30	182.4	40	182.4	120	1,642

Note: T = energy (MeV), S = mass collision stopping power for liquid water (MeVcm^{2}g^{-1}).

EXAMPLE 9.1

Calculate Pb collision stopping power for 1 GeV α-particles (assume $I = 823$ eV).

SOLUTION 9.1

$$\gamma = \frac{T + mc^2}{mc^2} = \frac{1 + 3.727}{3.727} = 1.268$$

$$\beta^2 = 1 - \gamma^{-2} = 1 - 1.268^{-2} = 0.3783$$

$$S_{col} = \frac{4\pi r_e^2 mc^2 z^2 Z \rho N_A}{\beta^2 A} \left[\ln \frac{2mc^2\beta^2}{I(1-\beta^2)} - \beta^2 \right]$$

$$= \frac{4\pi \times (2.82 \text{ fm})^2 \times 0.511 \text{ MeV} \times 2^2 \times 82 \times 11.34 \text{ g/cm}^3 \times 6.022 \times 10^{23}}{0.3783 \times 207 \text{ g/mol}}$$

$$\times (10^6 \text{ cm}^3/\text{m}^3) \left[\ln \frac{2 \times 1 \times 10^9 \times 0.3783}{823 \text{ eV}(1-0.3783)} - 0.3783 \right]$$

$$= 20.7 \text{ GeV/m}$$

EXAMPLE 9.2

Calculate the maximum energy loss at a collision with a free electron for 1 GeV α-particles.

SOLUTION 9.2

$$Q_{max} = \frac{2mc^2\beta^2}{1-\beta^2} \frac{1}{1+\frac{2m}{M\sqrt{1-\beta^2}}+\left(\frac{m}{M}\right)^2}$$

$$= \frac{2\times 0.511\ \text{MeV} \times 0.3783}{1-0.3787} \frac{1}{1+\frac{2\times 0.511\ \text{MeV}}{3727\ \text{MeV}\sqrt{1-0.3783}}+\left(\frac{0.511\ \text{MeV}}{3727\ \text{MeV}}\right)^2}$$

$$= 0.622\ \text{MeV}$$

9.2 Nuclear Stopping Powers

Energy loss due to the elastic scattering of heavy charged particles with a nucleus in the energy range lower than 10 keV cannot be ignored. This is called nuclear stopping powers, which means the recoil energy of the target nucleus derived from the kinematical relationship. Putting the differential cross section for elastic scattering as $d\sigma_{el}/d\Omega$, the mass nuclear stopping power S_{nuc}/ρ is represented by

$$\frac{S_{nuc}}{\rho} = \frac{2\pi N_A}{A} \int_0^\pi \frac{d\sigma_{el}}{d\Omega}\ W(\theta, T)\sin\theta d\theta \tag{9.11}$$

where $W(\theta, T)$ is the recoil energy, which depends on the scattering angle, θ, and the kinetic energy of the charged particle, T. Using the mass of the particle, M, and the mass of the target atom, M_t, $W(\theta, T)$ is given by

$$W(\theta, T) = 4T\frac{M_t M}{(M_t + M)^2}\ \sin^2\frac{\theta}{2} \tag{9.12}$$

The elastic scattering cross section for a high-energy region above a few MeV is described by the Rutherford cross section:

$$\frac{d\sigma_{el}}{d\Omega} = \frac{N_A Z^2 z^2 r_e^2}{4A}\left(\frac{mc}{p\beta}\right)^2 \frac{1}{\sin^4\frac{\theta}{2}} \tag{9.13}$$

in which the momentum of the heavy charged particle, p, is given by

$$p = \frac{1}{c}\sqrt{(T + Mc^2)^2 - (Mc^2)^2} \tag{9.14}$$

Since the radiative stopping power can be neglected for heavy charged particles, the total stopping power is the sum of collision loss and nuclear loss.

$$\frac{S_{tot}}{\rho} = \frac{S_{col}}{\rho} + \frac{S_{nuc}}{\rho} \tag{9.15}$$

The contribution of nuclear stopping power to the total stopping power is negligible for the energy region greater than 10 keV; however, it increases as the energy is lowered. That amounts to about 30% at 1 keV.

EXAMPLE 9.3

1. What is the probability of carbon ions scattering through 45°, given that the density of lead is 11.3 g/cm³?
2. What is the counting rate of carbon ions scattered through this angle, given that the flux, Ψ, of incident ions is 10^{11} per second on the foil? The lead target is 0.5 mm thick and a detector of detecting area, Δa, 1 cm² is placed 150 cm away from the collision center.
3. What is the energy loss of the incident ions in such a collision?

SOLUTION 9.3

1. The probability per unit thickness of the target is

$$P(\theta) = 2\pi \sin\theta \cdot \frac{\rho N_A}{A} \cdot \frac{d\sigma_{el}}{d\Omega} \quad (\text{unit } sr^{-1} \, cm^{-1})$$

The differential cross section is calculated from

$$\frac{d\sigma_{el}(45°)}{d\Omega} = \frac{6^2 \cdot 82^2 \cdot (2.818 \cdot 10^{-17} \, cm)^2}{4} \left(\frac{0.511 \, MeV}{pc\beta}\right)^2 \frac{1}{\sin^4 \frac{45°}{2}}$$

$$= \frac{5.851 \cdot 10^{-28}}{(pc\beta)^2} \, MeV^2 \, cm^{-1} \, sr^{-1}$$

from Equation (9.14) the relative velocity can be calculated:

$$\beta = \sqrt{1 - \left(\frac{M_C c^2}{T + M_C c^2}\right)^2} = \sqrt{1 - \left(\frac{12 \cdot 931.494 \, MeV}{1 \, MeV + 12 \cdot 931.494 \, MeV}\right)^2} = 0.013$$

where the momentum of the carbon ion is given by

$$pc = \sqrt{(T + M_C c^2) - (M_C c^2)^2}$$

$$= \sqrt{(1 \, MeV + 12 \cdot 931.494 \, MeV)^2 - (12 \cdot 931.494 \, MeV)^2}$$

$$= 149.52 \, MeV$$

therefore,

$$P(45°) = 2\pi \sin 45° \cdot \frac{(11.3\,\text{g/cm}^3) \cdot (6.022 \cdot 10^{23}\,\text{mol}^{-1})}{207.2\,\text{g/mol}}$$

$$\cdot \frac{5.851 \cdot 10^{-28}\,\text{MeV}^2\,\text{cm}^{-1}}{(0.013 \cdot 149.52\,\text{MeV})^2}$$

$$= 2.26 \cdot 10^{-5}\,\text{cm}^{-1}\,\text{sr}^{-1}$$

2. From the relationship $\Delta\Omega = 2\pi \sin\theta \cdot \Delta\theta = \dfrac{\Delta a}{r^2}$, the rate of scattered ions are calculated from

$$R(45°) = \Psi \cdot P(45°) \cdot \Delta\theta \cdot t = \frac{P(45°)}{2\pi \sin\theta} \cdot \frac{\Delta a}{r^2} \cdot t$$

$$= 10^{11}\,\text{s}^{-1} \cdot \frac{2.26 \cdot 10^{-5}\,\text{cm}^{-1}\,\text{sr}^{-1}}{2\pi \sin 45°} \cdot \frac{1}{(150)^2}\,\text{sr} \cdot (0.05\,\text{cm})$$

$$= 1.13\,\text{s}^{-1}$$

where t is the lead's thickness and $R(45°)$ is the counting rate at 45°.

3. $W = 4T \dfrac{M_C M_{Pb}}{(M_C + M_{Pb})^2} \cdot \sin^2 \dfrac{\theta}{2}$

$$= 4 \cdot 1\,\text{MeV} \frac{12 \cdot 207.2}{(12 + 207.2)^2} \cdot \sin^2 \frac{45}{2} = 0.2\,\text{MeV} \cdot 0.146 = 0.029\,\text{MeV}$$

2.9% of the incident energy.

In the low-energy region below 100 keV, Equation (9.13) cannot be used because of the increase of the screening effect due to the orbital electrons of an atom. A calculation method taking into account the screening effect is needed. In that case, the interaction of the incident particle with the target is represented by the screened Coulomb potential such as

$$V(r) = \frac{zZe^2}{r} F_s(r/r_s) \tag{9.16}$$

in which $F_s(r/r_s)$ is a function taking into account the screening, and r_s is a parameter representing the degree of screening. The differential cross section is obtained using the classical mechanics trajectory calculation.

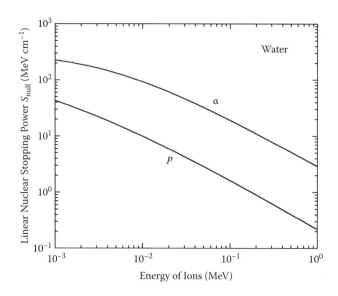

FIGURE 9.3
Linear nuclear stopping powers of water for protons and α-particles.

The solution is obtainable not analytically but numerically. Figure 9.3 shows the nuclear stopping powers of water for protons and α-particles.

9.3 Ranges

For heavy charged particles, unlike electrons, the effect of multiple scattering is not considered. It is approximated those continuously slow down nearly on a straight line. Originally, the continuous slowing down approximation (CSDA) was defined for heavy particles. The CSDA range $R(T)$ is obtained by

$$R(T) = \int_0^T \left[S_{col}(T) + S_{nuc}(T) \right]^{-1} dT \tag{9.17}$$

Figure 9.4 shows the ranges, in cm, of protons in various materials. Figure 9.5 shows the ranges of α-particles in air ($\rho = 1.205 \times 10^{-3}$ g cm^{-3}) under the standard condition. If the range of a kind of particle is known, the range of another kind of particle can be obtained. A relationship between

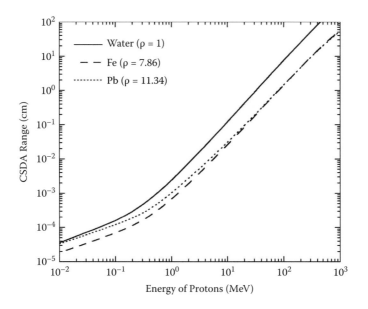

FIGURE 9.4
CSDA ranges of protons for various matters.

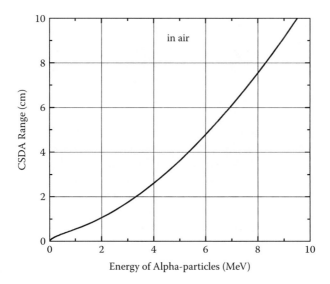

FIGURE 9.5
CSDA ranges of α-particles in air.

the ranges of different types of two particles with the same initial velocity is represented by

$$\frac{R_1(\beta)}{R_2(\beta)} = \frac{z_2^2 M_1}{z_1^2 M_2} \qquad (9.18)$$

Assuming the range of the proton as particle 2 is known, the range R of the other particle becomes

$$R(\beta) = \frac{M}{z^2} R_p(\beta) \qquad (9.19)$$

EXAMPLE 9.4

Estimate the range of an 8 MeV α-particle in water using Figure 9.4 for the proton range.

SOLUTION 9.4

The kinetic energy of a proton with the same speed as an 8 MeV α-particle is 2 MeV from Equation (9.9). Therefore, $R_\alpha = \frac{4}{2^2} R_p = R_p$. Using Figure 9.4, $R_\alpha = 7.5 \times 10^{-3}$ cm is obtained.

9.4 Straggling of Energy Loss and Range

When charged particles pass through the material, statistical fluctuation in the number of collision and the energy loss arises. Consequently, many particles started under the same condition show fluctuation in (1) the energy after traveling a given depth and (2) the pass length until stopping. The former phenomenon is called the energy loss straggling and the latter the range straggling. The distribution of the energy straggling becomes a Gaussian shape for the case of a large number of collisions between the charged particles and the absorber atoms. In this case, the whole energy loss is much greater than the maximum energy loss in a single collision. On the contrary, the distribution becomes asymmetrical for the case of a small number of collisions; in other words, for a short track segment.

The range straggling for heavy charged particles can be measured using the absorber and the counter. A monochromatic energy beam hits the absorber consisting of a variable thickness. Figure 9.6 shows the counting rate as a function of the absorber thickness. At a thin thickness, the number of particles is kept constant until the particle range increase leads to energy degradation. Then, the count number decreases abruptly. The reason for the shape is due to the range straggling. The thickness of the absorber at half of the total count is defined as the mean range. The extrapolated range

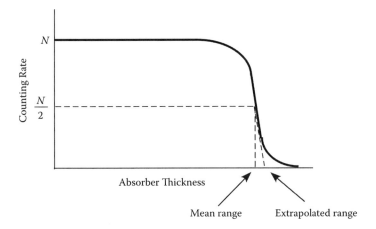

FIGURE 9.6
Measurement of heavy particle range (Uehara 2002).

is determined by the point of intersection of the extrapolated line with the abscissa. The range straggling is not large; for example, for 100 MeV protons in biological material it is ~1%. The ionization current is measured using an ionization chamber. The variation of the current per unit thickness as a function of the absorber depth is called the Bragg curve. The energy deposition near the end of the range enhances and forms the sharp peak called the Bragg peak, which is characteristic of heavy charged particles. Heavy ion cancer therapy utilizes this feature of heavy charged particles.

9.5 Summary

1. All particles with $26 \geq Z \geq 1$ are called heavy particles.

2. A heavy charged particle moving in matter affects the electromagnetic force on the atomic electron. The particle loses energy by ionizing and exciting the atom.

3. Collision stopping powers for heavy charged particles are evaluated using Bethe's formula. Nuclear stopping powers are evaluated using the classical mechanics trajectory calculations.

4. Total mass stopping power for heavy charged particles is the sum of mass collision stopping power and mass nuclear stopping power.

5. Heavy charged particles travel almost straightforward in matter. Ranges are obtained by the continuous slowing down approximation.

6. The energy deposition of heavy charged particles enhances just before stopping in matter. This phenomenon, called the Bragg peak, is utilized in the heavy ion cancer therapy.

QUESTIONS

1. What is the difference between the terms CSDA *range* and *penetration distance*?
2. On average, what percentages of the energy lost by a heavy ion track are by primary particle and d electrons?
3. Approximately, in a proton therapy beam, how much of the energy lost is in the Bragg peak area?
4. What is the difference between the terms *linear stopping power* and *linear energy transfer*?
5. What is the current definition of the term *LET*?

References

ICRU. 1993. *Stopping Powers and Ranges for Protons and Alpha Particles*. ICRU Report 37.
Uehara S. 2002. *Radiation Physics*, 4th ed. Tokyo: Nanzando Publishing (in Japanese).

For Further Reading

Turner JE. 1995. *Atoms, Radiation, and Radiation Protection*, 2nd ed. New York: John Wiley & Sons.

10

δ-Ray, Restricted Stopping Power, and LET

10.1 δ-Ray

Electrons and heavy charged particles moving in materials often cause emission of secondary electrons having enough energy to separate from the path of the initial particle. These electrons, called the δ-rays, make distinct track branches themselves. Figure 10.1 shows the projected tracks of a 10 MeV proton in water generated by Monte Carlo simulation. Some δ-rays emitted from the path of a proton are seen. In the figure, almost all secondary electrons are ejected vertically in the direction from the path of the heavy particles. This is explained from the kinematical relationship of the collision between a heavy particle and a free electron. The emission angle is given by

$$\cos\theta = \sqrt{\frac{m_{HI}\varepsilon}{4m_eT}} \tag{10.1}$$

EXAMPLE 10.1

What is the energy of electrons emitted from a 12 MeV carbon ion path through an angle 45° relative to the projectile's path?

SOLUTION 10.1

From Equation (10.1):

$$\varepsilon = \frac{4m_eT}{m_{HI}}\cos^2\theta = \frac{4\cdot 5.486\cdot 10^{-4}u\cdot 12\ \text{MeV}}{12\ u}\cos^2 45° = 1.097\ \text{keV}$$

---

in which m_{HI} and T are the mass and kinetic energy for a heavy particle, and m_e and ε for an electron, respectively. For 1 MeV protons, the average energy of secondary electrons is ~60 eV. Therefore, $\theta \sim 80°$ is obtained. This is understood by the momentum conservation law. The momentum of heavy particles does not change largely before and after the collision; therefore, the momentum component along the particle path for secondary electrons becomes nearly zero. There is not an accurate criterion to classify δ-rays.

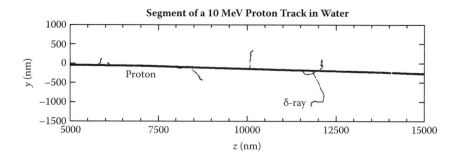

FIGURE 10.1
δ-Rays generated by a 10 MeV proton in water. Dots represent the energy deposition points.

The absorbed dose of radiation is defined as the energy absorbed per unit mass of irradiated material. The stopping power means the energy lost by charged particles in the material. This quantity is not necessarily equal to the absorbed energy by the target. Above all, a problem occurs for the target smaller than the range of secondary electrons. Such a microscopic concept in dosimetry, called microdosimetry, plays an important role to understand the radiation effect on cells (~μm range) or DNA (at ~nm range).

10.2 Restricted Stopping Power

The concept of restricted stopping power was introduced to relate the energy loss in the target with that actually absorbed energy in the target. The restricted collision stopping power, $-(dE/dx)_\Delta$, is defined as the energy loss just by a collision in which the energy transfer does not exceed a value Δ, which is called the cutoff. Replacing Q_{max} in Equation (7.3) by Δ,

$$-\left(\frac{dE}{dx}\right)_\Delta = \mu \int_{Q_{min}}^{\Delta} QW(Q)dQ \tag{10.2}$$

is obtained. The explicit formula of mass restricted collision stopping power for heavy charged particles is given by

$$-\left(\frac{dE}{\rho dx}\right)_\Delta = \frac{2\pi r_e^2 mc^2 z^2 ZN_A}{\beta^2 A}\left[\ln\frac{2mc^2\beta^2\Delta}{I^2(1-\beta^2)} - \frac{(1-\beta^2)\Delta}{2mc^2} - \beta^2 - 2\frac{C}{Z} - \delta\right] \tag{10.3}$$

For electrons,

$$-\left(\frac{dE}{\rho dx}\right)_\Delta = \frac{2\pi r_e^2 mc^2 z^2 ZN_A}{\beta^2 A}\left[\ln\frac{T^2}{I^2} + \ln\left(1+\frac{\tau}{2}\right) + G - \delta\right] \tag{10.4}$$

TABLE 10.1

Restricted Mass Stopping Powers of Water for Protons
(MeV cm^2 g^{-1})

Energy (MeV)	$-\left(\dfrac{dE}{\rho dx}\right)_{100\,eV}$	$-\left(\dfrac{dE}{\rho dx}\right)_{1\,keV}$	$-\left(\dfrac{dE}{\rho dx}\right)_{10\,keV}$	$-\left(\dfrac{dE}{\rho dx}\right)_{\infty}$
0.05	910.0	910.0	910.0	910.0
0.10	711.0	910.0	910.0	910.0
0.50	249.0	424.0	428.0	428.0
1.0	146.0	238.0	270.0	270.0
10.0	24.8	33.5	42.2	45.9
100.0	3.92	4.94	5.97	7.28

where $\tau = T/mc^2$, $\eta = \Delta/T$, and G is a function of τ and η as follows:

$$G = -1 - \beta^2 + \ln[4\eta(1-\eta)] + \frac{1}{1-\eta} + (1-\beta^2)\left[\frac{\tau^2\eta^2}{2} + (2\tau+1)\ln(1-\eta)\right] \quad (10.5)$$

Table 10.1 shows the mass restricted collision stopping powers of water for various Δ for protons. Unrestricted stopping power is represented by $\Delta = \infty$. For protons below 0.05 MeV, collision transfer energy greater than 100 eV does not contribute to the stopping power. In fact, Q_{max} is 109 eV at 0.05 MeV protons. For 0.1 MeV protons, the restricted stopping power at $\Delta = 1$ keV is much greater than that at $\Delta = 100$ eV because $Q_{max} = 220$ eV. It is found that the energy transfer greater than 10 keV is scarce at 1 MeV. For 10 MeV protons, about 8% of the stopping power comes from the transfers greater than 10 keV. Table 10.2 shows the mass restricted collision stopping powers of water for electrons.

The value of Δ is chosen in accordance with the objective of the subject. We refer Monte Carlo simulation of electron transport as an example. If $-(dE/dx)_{1eV}$ is assumed, the secondary electron having an energy greater than 1 keV is treated as independent electron tracks or δ-rays. The secondary electrons having smaller energies than 1 keV are lumped in the restricted collision stopping powers. The restricted radiative stopping power is treated in the same manner. Bremsstrahlung photon energy less than Δ is included in the restricted radiative stopping power. On the other hand, photons having energies higher than Δ are traced as individual photons.

EXAMPLE 10.2

Estimate the dose deposited by a plane-parallel monoenergetic electron beam of 1 MeV in a water target that has a cylindrical shape with the length and radius equal to the range of 10 keV electrons. The cylindrical axis is centered on the particle's path and the electron fluence is 10^{10} cm^{-2}.

TABLE 10.2

Restricted Mass Stopping Powers of Water for Electrons (MeV cm² g⁻¹)

Energy (MeV)	$-\left(\dfrac{dE}{\rho dx}\right)_{100\,eV}$	$-\left(\dfrac{dE}{\rho dx}\right)_{1\,keV}$	$-\left(\dfrac{dE}{\rho dx}\right)_{10\,keV}$	$-\left(\dfrac{dE}{\rho dx}\right)_{100\,keV}$	$-\left(\dfrac{dE}{\rho dx}\right)_{\infty}$
0.01	15.3	20.2	23.2	23.2	23.2
0.02	8.63	11.2	13.5	13.5	13.5
0.03	6.22	8.02	9.62	9.88	9.88
0.05	4.18	5.32	6.38	6.75	6.75
0.07	3.25	4.12	4.94	5.31	5.31
0.1	2.54	3.19	3.82	4.20	4.20
0.2	1.69	2.09	2.49	2.84	2.84
0.3	1.41	1.73	2.05	2.35	2.39
0.5	1.19	1.46	1.72	1.97	2.06
0.7	1.11	1.34	1.58	1.81	1.94
1.0	1.05	1.27	1.49	1.71	1.86
2.0	1.02	1.22	1.42	1.63	1.84
3.0	1.02	1.22	1.42	1.62	1.87
5.0	1.03	1.22	1.42	1.62	1.92
7.0	1.03	1.23	1.43	1.63	1.95
10.0	1.04	1.24	1.44	1.63	1.99
20.0	1.05	1.25	1.44	1.64	2.06
30.0	1.05	1.25	1.45	1.64	2.10
50.0	1.05	1.25	1.45	1.64	2.14

SOLUTION 10.2

Dose can be estimated from

$$D \approx \phi \left(-\frac{dE}{\rho dx}\right)_{1\,keV} = \frac{10^{10}}{cm^2} \cdot \left(\frac{1.49 \cdot 10^6 \cdot 1.6 \cdot 10^{-19} \text{ J cm}^2}{10^{-3}\,\text{kg}}\right) = 2.38 \text{ Gy}$$

EXAMPLE 10.3

What should be the cutoff energy for the dose calculation for a cylindrical water target of radius 2.5 μm?

SOLUTION 10.3

2.5 μm is about the range of 10 keV electrons in liquid water. Therefore, the restricted stopping power with cutoff energy $\Delta \geq 10$ keV should be used for the dose calculation.

10.3 LET

LET stands for linear energy transfer. The concept of LET is defined by ICRU Report 16 (1970). It is denoted by L_Δ, which is the same as the restricted stopping power for the energy transfer less than Δ.

$$L_\Delta = \left(\frac{\mathrm{d}E}{\mathrm{d}l} \right)_\Delta \tag{10.6}$$

in which $\mathrm{d}l$ is the path length of the particle, and $\mathrm{d}E$ is the mean energy loss less than the cutoff energy Δ. The L_∞ means the ordinary stopping power. L without ∞ means L_∞. Although not included in the present definition, another LET, L_r, other than L_Δ has been considered that restricts the position of energy deposition. This quantity is defined as the energy deposited in a cylinder with the radius r and the length $\mathrm{d}l$ centered on the particle path.

$$L_r = \left(\frac{\mathrm{d}E}{\mathrm{d}l} \right)_r \tag{10.7}$$

It is easy to evaluate L_Δ using the analytical equations ((10.3) and (10.4)); however, it is difficult to measure it directly. Figure 10.2 shows the ratio of

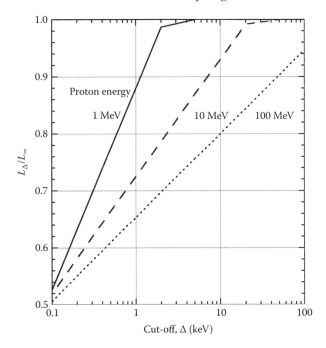

FIGURE 10.2
Ratio of energy restricted stopping power L_Δ to unrestricted stopping power L_∞ for the proton energies 1, 10, and 100 MeV as a function of cutoff energy D.

L_Δ to L_∞ as a function of Δ for various energies of protons. Analytical calculations are difficult for L_r, but the measurement is basically possible by putting the cylindrical gas ionization chamber in place. Further analyses are given in Section 18.5.

10.4 Summary

1. Some of the secondary electrons generated by ionization are energetic. Those called δ-rays form an independent track of the primary electron.

2. The concept of macroscopic and averaged stopping powers cannot be applied to the absorbed dose in the target the size of μm or nm.

3. Restricted collision stopping powers are defined as the energy transfer caused collision not exceeding the cutoff Δ.

4. Linear energy transfer (LET) is the same as the restricted stopping power. The notation is L_Δ.

QUESTIONS

1. What is the condition for normal ejection of secondary electrons from a high-energy ion's path?

2. Explain why the restricted stopping powers for a 50 keV proton are the same for the cutoff energy $\Delta = 100$ eV, 1 keV, 10 keV, and ∞. Why does the difference become larger when the proton energy increases?

3. What is the condition of using $\left(\frac{dE}{\rho dx}\right)_\Delta$ for dose calculations in a target volume?

4. Give reasons why absorbed dose calculated using the approach given in Question 3 deviates from the measured value.

References

ICRU. 1970. *Linear Energy Transfer*. ICRU Report 16.
ICRU. 1983. *Microdosimetry*. ICRU Report 36.
ICRU. 1984. *Stopping Powers for Electrons and Positrons*. ICRU Report 37.

11

Introduction to Monte Carlo Simulation

11.1 Monte Carlo Method

The Monte Carlo method is one of the simulation techniques in which the computer experiment is done generating pseudorandom numbers repeatedly. If the probability law for the elementary processes of the relevant phenomenon is known, these processes are generated on the computer as if they actually occur and the behavior of the whole system is investigated. Thus, Monte Carlo simulation is able to reproduce faithfully the phenomenon without any approximations. This is the most accurate calculation method in order to examine the complex transport phenomenon of radiation in matter. Figure 11.1 shows microscopic tracks of five 10 keV electrons in water calculated by Monte Carlo simulation. Because of the stochastic nature of radiation, each track is not identical. Therefore, a lot of trials are accumulated in order to obtain meaningful results.

Interactions of radiations such as photons, electrons, and ions with matter are almost known, and quantitative data such as cross sections, energy transfers, energy spectra, and angular distributions are available for each interaction. A computer program is constructed combining these fundamental data. In this chapter, some basic knowledge of Monte Carlo programming for photons and electrons is described.

11.2 Sampling of Reaction Point

The probability of a first collision between x and $x + dx$ along its line of flight is given by

$$p(x)dx = \mu(x)dx\, e^{-\int_0^x \mu(s)ds} \tag{11.1}$$

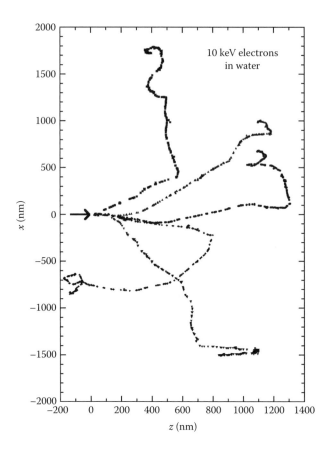

FIGURE 11.1
Tracks of five 10 keV electrons in water.

where $\mu(x)$ is the macroscopic cross section or the inverse mean free path. For photons, this is equal to the linear attenuation coefficient. The cumulative probability distribution function P is obtained by integration,

$$P = 1 - e^{-N(x)} \qquad (11.2)$$

where $N(x)$ is

$$N(x) = \int_0^x \mu(s)\,ds \qquad (11.3)$$

The function P or $1 - P$ corresponds to the uniform random number [0, 1]. These equations are applicable not only to uncharged particles but also to charged particles.

EXAMPLE 11.1

A photon beam is traveling in a uniform infinite material with the attenuation coefficient μ. What is the probability and differential probability of interaction at a distance x?

SOLUTION 11.1

Attenuation law, $p(x) = e^{-\mu x}$

$P(x) = 1 - p(x) = 1 - e^{-\mu x}$, probability of interaction at a distance x

$\dfrac{dP(x)}{dx} = \mu e^{-\mu x} = \mu p(x)$

Particles travel freely until the next reaction; therefore, $N(x) = \mu x$ because of $\mu = $ constant. If P is replaced by $1 - r$ for a random number r, the free path length x is given by a famous formula:

$$x = -\frac{\ln r}{\mu} \tag{11.4}$$

EXAMPLE 11.2

Calculate the path length for photons with $\mu = 0.1$ cm^{-1} as a function of random numbers between 0 and 1.

SOLUTION 11.2

The free path length, x, is obtained using Equation (11.4).

Random Number r	ln r	x (cm)
0.1	−2.303	23.03
0.2	−1.609	16.09
0.3	−1.204	12.04
0.4	−0.916	9.16
0.5	−0.693	6.93
0.6	−0.511	5.11
0.7	−0.357	3.57
0.8	−0.223	2.23
0.9	−0.105	1.05

If the unit of μ is cm^{-1}, then the free path length is given by cm units. Generally, this formula is used for photons and neutrons. However, this

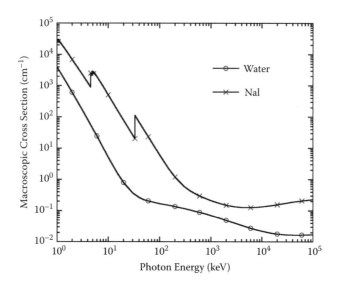

FIGURE 11.2
Macroscopic cross sections for photons in water and NaI.

equation is also applicable to both charged and uncharged particles such as electrons and heavy ions if the cross sections for individual interactions are available. The cross sections for charged particles used for Monte Carlo simulation of microscopic track structure are given in Chapters 12–16.

Using the atomic cross section for photons, μ is calculated by

$$\mu = \rho \frac{N_A}{M} \sum_i n_i \sigma_i^{tot} \tag{11.5}$$

in which M is the mass of compound, and n_i and σ_i^{tot} are the number and the total cross section (cm^2) for the ith atom forming the compound. Figure 11.2 shows the macroscopic cross section μ (cm^{-1}) for photons in water and NaI.

EXAMPLE 11.3

Calculate the linear attenuation coefficient of photons for NaI in the energy range between 10 keV and 1 MeV.

SOLUTION 11.3

Use $\rho = 3.667$ g cm^{-3}, $M = 149.88$ g, $\rho \frac{N_A}{M} = 1.4734 \times 10^{22}$.

Energy (MeV)	σ_{Na}^{tot} (b)	σ_I^{tot} (b)	μ (cm^{-1})
0.01	579.12	33,803	506.58
0.02	76.9	5,265.3	78.71
0.03	27.16	1,778	26.60
0.03317	23.52	1,376.5	20.63
0.03318	23.42	7,565.6	111.82
0.04	14.99	4,634.4	68.50
0.05	10.63	2,575.4	38.10
0.06	8.616	1,587.3	23.51
0.08	6.849	732.1	10.89
0.1	6.030	403.8	6.038
0.2	4.571	76.45	1.194
0.3	3.927	36.94	0.6021
0.4	3.501	25.44	0.4264
0.5	3.189	20.37	0.3471
0.6	2.953	17.44	0.3004
0.8	2.586	14.16	0.2467
1.0	2.324	12.204	0.2141

11.3 Condensed History Technique

The electron energy region of interest in radiation diagnosis and radiation therapy range from about 10 keV to 20 MeV. The method that calculates the free path length event by event becomes impractical because this method requires a huge computation time and computer memory as the radiation energy increases. Berger (1963) devised a treatment procedure called the condensed history technique for charged particles. Here, we explain the condensed history technique referring to electrons (Nelson et al. 1985; Uehara 1986). This approximates the electron degradation by multiple scattering theories in combination with a small number of discrete events and uses the restricted stopping power due to ionization and excitation processes to describe the slowing down of electrons. The discrete process involves the following interactions:

1. Electron-induced bremsstrahlung and inelastic scattering (Møller scattering)
2. Positron-induced bremsstrahlung, inelastic scattering, (Bhabha scattering) and annihilation in flight

Secondary electrons are generated by Møller scattering and Bhabha scattering. Secondary photons are generated by bremsstrahlung and annihilation in flight. The cross section formula for bremsstrahlung is described in Section 12.5. The differential cross sections in energy for electron-electron inelastic scattering (Møller scattering) are given by

$$\frac{d\sigma}{dw} = \frac{2\pi Z r_e^2}{T^2 \beta^2} \left[\frac{1}{\varepsilon^2} + \frac{1}{(1-\varepsilon)^2} + \frac{T^2}{(1+T)^2} - \frac{1}{\varepsilon(1-\varepsilon)} \frac{2T+1}{(1+T)^2} \right] \tag{11.6}$$

in which T represents the initial energy of the electron and w represents the energy of a secondary electron. Both are in mc^2 units. ε is the ratio w/T. Because the primary electron and the orbital electron cannot be distinguished, the higher-energy electron is regarded as the primary electron. The total cross section for the secondary electrons having energy greater than the cutoff Δ is obtained by integrating the above equation in the interval $[\Delta, T/2]$.

$$\sigma = \frac{2\pi Z r_e^2}{T \beta^2} \left[\frac{T}{\Delta} - \frac{T}{T-\Delta} + \frac{T^2}{(1+T)^2} \left(\frac{1}{2} - \frac{\Delta}{T} \right) - \frac{2T+1}{(1+T)^2} \ln \frac{T-\Delta}{\Delta} \right] \tag{11.7}$$

The differential cross sections in energy for positron-electron inelastic scattering (Bhabha scattering) are given by

$$\frac{d\sigma}{dw} = \frac{2\pi Z r_e^2}{T^2} \left[\frac{1}{\varepsilon} \left(\frac{1}{\varepsilon \beta^2} - B_1 \right) + B_2 + \varepsilon(\varepsilon B_4 - B_3) \right] \tag{11.8}$$

in which

$$y = \frac{1}{2+T}, \qquad B_1 = 2 - y^2, \qquad B_2 = (1-2y)(3+y^2),$$

$$B_4 = (1-2y)^3, \qquad B_3 = B_4 + (1-2y)^3 \tag{11.9}$$

The total cross section for the secondary electrons having energy greater than the cutoff Δ is obtained by integrating the above equation in the interval $[\Delta, T]$.

$$\sigma = \frac{2\pi Z r_e^2}{T} \left[\frac{1}{\beta^2} \left(\frac{T}{\Delta} - 1 \right) - B_1 \ln \frac{T}{\Delta} + B_2 \left(1 - \frac{\Delta}{T} \right) + \left(\frac{B_4}{3} - \frac{B_3}{2} \right) - \frac{\Delta^2}{T^2} \left(\frac{B_4 \Delta}{3T} - \frac{B_3}{2} \right) \right] \tag{11.10}$$

The differential cross sections in energy for annihilation in flight are given by

$$\frac{d\sigma}{dk} = S_1(k) + S_1(A-k) \tag{11.11}$$

in which k is the photon energy and

$$E = T + 1, \qquad A = E + 1, \qquad p = \sqrt{E^2 - 1}$$

$$S_1(x) = C_1\left(-1 + \frac{C_2 - 1/x}{x}\right), \qquad C_1 = \frac{\pi Z r_e^2}{AT}, \qquad C_2 = A + \frac{2E}{A} \qquad (11.12)$$

The energy of the other annihilation quanta becomes $T + 2 - k$ in mc^2 units. The total cross section is given by

$$\sigma = \frac{\pi Z r_e^2}{E + 1}\left[\frac{E^2 + 4E + 1}{p^2}\ln(E + p) - \frac{E + 3}{p}\right] \qquad (11.13)$$

Since electrons lose energy almost continuously, the value of μ varies during the slowdown, unlike photons. If an electron with the kinetic energy T_0 and the total energy $E_0 = T_0 + mc^2$ degrades to T_1 and $E_1 = T_1 + mc^2$, respectively, during the path length x, $N(x)$ for all discrete processes is modified as follows:

$$N(x) = \sum_i N_i(x) \qquad (11.14)$$

in which $N_i(x)$ is for the ith process,

$$N_i(x) = \int_{E_0}^{E_1} \frac{\mu_i(E)}{(-dE/dx)_{\text{conti}}}\, dE \equiv f_i(T_0) - f_i(T_1) \qquad (11.15)$$

A dimensionless function $f(T)$ means the number of secondary particles generated until the electron with T stops. This corresponds to μx for photons. The denominator is the restricted stopping power; in other words, the continuous stopping power.

$$(-dE/dx)_{\text{conti}} = (-dE/dx)_{\text{col},\Delta} + (-dE/dx)_{\text{rad},\Delta} \qquad (11.16)$$

The values of cutoff Δ for collision and radiative stopping powers are set independently. The value of $f_i(T)$ is numerically evaluated using the total cross sections σ_i for the ith process assuming the number of the jth atom contained in a 1 cm^3 molecule, n_j.

$$f_i(T) = \sum_j n_j \int_{E_{th}}^{T+1} \frac{\sigma_i(E)}{(-dE/dx)_{\text{conti}}}\, dE \qquad (11.17)$$

Figure 11.3 shows an example of calculated $f(T)$ values of four interactions for NaI.

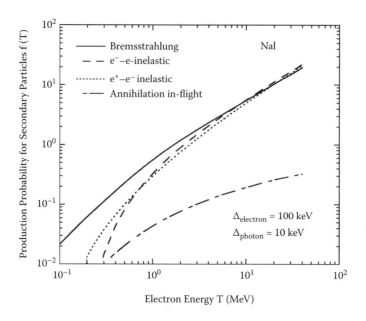

FIGURE 11.3
Production probabilities for secondary particles, $f(T)$.

The caption [100, 10] in the figure means the cutoff for electrons and photons, respectively, in keV units. From Equations (11.2) and (11.15),

$$P = 1 - e^{f(T_1) - f(T_0)} \tag{11.18}$$

is obtained. If $T_1 \sim 0$, then $f(T_1) \sim 0$. Then, P becomes

$$P = 1 - e^{-f(T_0)} \tag{11.19}$$

P is compared with a random number, r. If $r > P$, then the electron becomes a continuous process. If not, one of the reactions occurs. The energy T_1, at which reaction occurs, is determined from Equation (11.18),

$$f(T_1) = f(T_0) + \ln(1 - r') \tag{11.20}$$

where r' means a new random number. Thus, the energy at which interaction occurs is randomly sampled by a table lookup method using the predigested $f(T)$ table. The type of reaction is selected in proportion to the magnitude of $f_i(T_0)$. The condensed history technique lumps individual events of energy deposition less than Δ into the continuous stopping power and regards the high-energy secondary particle often emitted as a new particle. This technique is called mixed procedure.

EXAMPLE 11.4

Calculate $f(T)$ values for four interactions under the condition $T = 1$ MeV electrons in Pb, assuming $[\Delta_e, \Delta_p] = [100, 10]$ keV.

SOLUTION 11.4

Constant parameters are $\rho = 11.34$ g cm^{-3}, $A = 207.2$ g, and $(-dE/dx)_{rad,10keV} = 44.19$ keV cm^2 g^{-1}. Both total cross sections σ and the continuous stopping power $(-dE/dx)_{conti}$ are represented as a function of $E = T + mc^2$. Integration is done using the mathematical subroutine installed on the computer.

1. Bremsstrahlung. The analytical form for the total cross section is unavailable; the double integral is needed for the bremsstrahlung. The differential cross section $d\sigma/dk = A_E f_E \, d\sigma^{3BN}/dk$ is adopted for $T = 1$ MeV (Equation 12.15).

$$f_{Br}(T) = \rho \frac{N_A}{A} \int_{\Delta_e + mc^2}^{T + mc^2} \frac{\int_{\Delta_p}^{E - mc^2} A_E f_E \dfrac{d\sigma^{3BN}}{dk} dk}{(-dE/dx)_{conti}} dE = 1.081$$

2. Møller scattering. The maximum energy of the secondary electron is $T/2$; therefore, the lower boundary of E becomes $2\Delta_e + mc^2$ if the cutoff is Δ_e.

$$f_{M\varphi}(T) = \rho \frac{N_A}{A} \int_{2\Delta_e + mc^2}^{T + mc^2} \frac{\sigma_{M\varphi}}{(-dE/dx)_{conti}} dE = 0.350$$

3. Bhabha scattering.

$$f_{Bh}(T) = \rho \frac{N_A}{A} \int_{\Delta_e + mc^2}^{T + mc^2} \frac{\sigma_{Bh}}{(-dE/dx)_{conti}} dE = 0.317$$

4. In-flight annihilation.

$$f_{Ia}(T) = \rho \frac{N_A}{A} \int_{\Delta_e + mc^2}^{T + mc^2} \frac{\sigma_{Ia}}{(-dE/dx)_{conti}} dE = 0.047$$

The sum of these values becomes $f(1 \text{ MeV}) = 1.081 + 0.350 + 0.317 + 0.047 = 1.795$. The table of $f_i(T)$ and $f(T)$ is prepared for various T. Using Equation (11.20), the energy T_1 at which the next interaction occurs is randomly sampled by a table lookup method.

11.4 Slowing Down of Electrons

In the continuous slowing down model, the electron path is artificially divided into a large number of short segments in such a way that the deposited energy per step is a certain fraction of the electron kinetic energy. Here the fraction s, which is called the step size, was kept constant during the slowing down. It is assumed that an electron of the kinetic energy of T_i enters the ith segment in which the energy deposition is ΔT_i. Using the step size s, ΔT_i is equal to sT_i. The corresponding thickness of the segment, t_i, is calculated by

$$t_i = \frac{\Delta T_i}{(-dE/dx)_{\text{conti}}} \tag{11.21}$$

EXAMPLE 11.5

Calculate the thickness of the segment in water for various electron energies assuming $[\Delta_e, \Delta_p] = [10, 10]$ keV and $s = 2\%$.

SOLUTION 11.5

$(-dE/dx)_{\text{rad,10keV}} = 5.129 \times 10^{-3}$ MeV cm^2 g^{-1}

T (MeV)	ΔT (MeV)	$(-dE/dx)_{\text{conti}}$ (MeV cm^{-1})	t (cm)
0.1	0.002	3.820	5.24×10^{-4}
1	0.02	1.494	1.34×10^{-2}
5	0.1	1.423	7.03×10^{-2}
10	0.2	1.436	1.39×10^{-1}
20	0.4	1.444	2.77×10^{-1}

This t_i corresponds to the thickness t introduced in the multiple scattering theory for electrons described in Section 8.5. The thickness becomes shorter gradually because T_i degrades upon increasing i. The ordinary value of s is set at 2 ~ 4%; therefore, the full range of electron is divided into 100 ~ 200 segments. At the point that an electron enters the segment, the deflection angle due to multiple scattering is sampled and emission of bremsstrahlung or δ-ray is randomly judged. If any interaction does not occur, the electron goes to the next segment with the energy degraded by the product of $(dE/dx)_{\text{conti}} \times$ (segment thickness). Otherwise, the energy, and the point and direction of emitted secondary particle, are randomly sampled. These procedures are needed for every segment because the energy of electron degrades at the entrance of the next segment.

11.5 Conversion of Angles

The emission angle of photons and electrons generated by an interaction is given by the angle on the polar coordinate set the direction of the incident particle. That is the same for the deflection angle in multiple scattering, too. However, in the case of many interactions occuring frequently, the incident direction differs every time. Therefore, it is necessary to convert the angle on the interaction system into the angle observed from the coordinate system fixed in the laboratory. We call this system the laboratory system. Figure 11.4 shows a relationship between the laboratory system (x, y, z) and the interaction system (x', y', z'). It is assumed that a particle incident with the polar angle θ and the azimuth angle ϕ interacts at the point P. The direction cosine of the incident particle is denoted by (l_0, m_0, n_0).

$$l_0 = \sin\theta\cos\phi, \qquad m_0 = \sin\theta\sin\phi, \qquad n_0 = \cos\theta \qquad (11.22)$$

After collision, it is assumed a particle is ejected to the polar angle ψ and the azimuth angle ρ on the interaction system (x', y', z'). Its direction cosine (l', m', n') is represented by

$$l' = \sin\psi\cos\rho, \qquad m' = \sin\psi\sin\rho, \qquad n' = \cos\psi \qquad (11.23)$$

The problem is to convert (ψ, ρ) into the direction observed on the laboratory system. First, a new coordinate system (x'', y'', z'') is formed by rotating

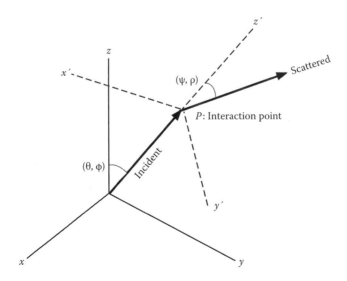

FIGURE 11.4
Relationship between the laboratory system (x, y, z) and the interaction system (x', y', z').

the (x', y', z') system around the y' axis. The direction cosine on this system becomes

$$
\begin{pmatrix} l'' \\ m'' \\ n'' \end{pmatrix} = \begin{pmatrix} \cos\theta & 0 & \sin\theta \\ 0 & 1 & 0 \\ -\sin\theta & 0 & \cos\theta \end{pmatrix} \begin{pmatrix} l' \\ m' \\ n' \end{pmatrix} \tag{11.24}
$$

Next, the (x'', y'', z'') system is rotated by $-\phi$ around the new z'' axis. Thus, the system returns to the (x, y, z) system. If the direction cosine on the (x, y, z) system is put as (l, m, n),

$$
\begin{pmatrix} l \\ m \\ n \end{pmatrix} = \begin{pmatrix} \cos\phi & -\sin\phi & 0 \\ \sin\phi & \cos\phi & 0 \\ 0 & 0 & 1 \end{pmatrix} \begin{pmatrix} l'' \\ m'' \\ n'' \end{pmatrix} \tag{11.25}
$$

is obtained. After all, (ψ, ρ) observed on the laboratory system becomes

$$
l = l_0 \cos\psi + \frac{1}{\sqrt{1-n_0^2}} \left(l_0 n_0 \sin\psi \cos\rho - m_0 \sin\psi \sin\rho \right)
$$

$$
m = m_0 \cos\psi + \frac{1}{\sqrt{1-n_0^2}} \left(m_0 n_0 \sin\psi \cos\rho + l_0 \sin\psi \sin\rho \right) \tag{11.26}
$$

$$
n = n_0 \cos\psi - \sqrt{1-n_0^2} \sin\psi \cos\rho
$$

These become the initial direction (l_0, m_0, n_0) on the laboratory system for the next reaction. If $n_0^2 \sim 1$ $(\theta \sim 0)$, then the same equations as (11.23) are obtained from Figure 11.4.

$$
l = \sin\psi \cos\rho, \qquad m = \sin\psi \sin\rho, \qquad n = \cos\psi \tag{11.27}
$$

11.6 Intersection at Boundary

When several different matters with certain geometrical shapes exist, the intersection at the boundary needs to be determined. Sampling of the free path is again done at that point. The form of the boundary surface is generally represented by a three-dimensional quadratic function:

$$
f(x, y, z) = 0 \tag{11.28}
$$

On the other hand, assumed is a particle started at (x_0, y_0, z_0) with the direction cosine (l, m, n). The equation of the straight line is represented by

$$
\frac{x - x_0}{l} = \frac{y - y_0}{m} = \frac{z - z_0}{n} \tag{11.29}
$$

The intersection between the boundary surface and the particle path can be obtained by solving the above two equations. The solution of z is given by

$$z_\pm = \frac{B \pm \sqrt{B^2 - AC}}{A} \tag{11.30}$$

For example, the equation for a sphere of the radius a is

$$x^2 + y^2 + z^2 - a^2 = 0 \tag{11.31}$$

The variables A, B, and C in Equation (11.30) are given by

$$A = 1$$
$$B = (l^2 + m^2)z_0 - n(lx_0 + my_0)$$
$$C = (l^2 + m^2)z_0^2 - 2nz_0(lx_0 + my_0) + n^2\left(x_0^2 + y_0^2 - a^2\right) \tag{11.32}$$

If $B^2 < AC$, then there is no intersection. Otherwise, x and y are given by

$$x_\pm = x_0 + \frac{l}{n}(z_\pm - z_0), \qquad y_\pm = y_0 + \frac{m}{n}(z_\pm - z_0) \tag{11.33}$$

EXAMPLE 11.6

An example of electron-photon Monte Carlo simulation applied to medical physics: What is the percentage depth dose in water for a ^{60}Co photon beam? The parameters of calculation are SSD = 100 cm, field size = 2 cm diameter, step size s = 2%, and cutoff = [50, 50] keV.

SOLUTION 11.6 (SEE ALSO FIGURE 1.4a)

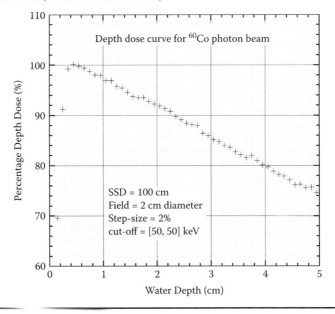

EXAMPLE 11.7

An example of the γ-camera imaging used in nuclear medicine: The figure shows the ^{111}In line source image located in the water phantom, which emits 245 keV photons.

SOLUTION 11.7

11.7 Summary

1. Monte Carlo simulation is a powerful tool to investigate the radiation transport phenomenon in matter.

2. The path length to the next reaction point, x, is sampled using the equation $x = -\ln r/\mu$, in which r is the uniform random number in the interval [0, 1] and μ is the macroscopic cross section.

3. The condensed history technique approximates the electron degradation by multiple scattering theories in combination with a small number of discrete events and uses the restricted stopping power due to ionization and excitation processes to describe the slowing down of electrons.

4. Electron interactions such as bremsstrahlung, Møller scattering, Bhabha scattering, and annihilation in flight are taken into account as the discrete events.

5. Conversion of the scattering angle in the interaction system to the angle in the laboratory system is needed to realize the measurement condition.

QUESTIONS

1. a. Write a simple computer program to select 100 random integer numbers between 1 and 10.
   b. Plot the frequency of the selected numbers.
   c. How uniformly are these numbers distributed?

2. Similar to question 1, select 1,000 random numbers between 0.0 and 1.0, and plot the frequency of occurrence of the numbers. How uniformly are these numbers distributed?

3. If the dose to a medium from photon irradiation, under the electronic equilibrium, can be calculated from the equation Dose$(r) = \iint E.\Phi(r,\Omega,E) (\mu_{en}/\rho)d\Omega \, dE$, define the elements of the equation. What are the conditions for small and extended boundary situations?

4. a. Write a computer program to select random points in a circle with radius r centered at (0,0).
   b. Write a computer program to select random m points in a sphere of radius r positioned at (0,0,0).

   What method is used to test uniform distribution of points in (a) and (b)? Design a flowchart for computer simulation of photon transport in a medium by Monte Carlo methods.

5. Write the formula to calculate the error and variance of a Monte Carlo calculation. In order to calculate the π number, one can sample the number of points inside a sphere, which is in the middle of a square. Write the program to calculate the π number. Calculate the error for 10, 100, and 1,000 samples.

References

Berger MJ. 1963. Monte Carlo calculations of the penetration and diffusion of fast charged particles. *Methods Comput. Phys.* 1: 135–215.

Nelson WR, Hirayama H, Rogers DWO. 1985. *The EGS4 Code System.* SLAC-Report-265.

Uehara S. 1986. The development of a Monte Carlo code simulating electron-photon showers and its evaluation by various transport benchmarks. *Nucl. Instrum. Methods* B14: 559–570.

Section II

12

Cross Sections for Interactions of Photons with Matter

12.1 Coherent Scattering

The differential cross section of Thomson scattering is represented by

$$\frac{d\sigma_T}{d\Omega} = \frac{r_e^2}{2}(1 + \cos^2\theta) \tag{12.1}$$

in which $r_e = 2.81794 \times 10^{-13}$ cm is the classical electron radius. This formula provides the cross section of an electron. From the formula, it is understood that angular distributions in Thomson scattering are independent of both the energy of incident photons and the target atom. The total cross section is calculated by

$$\sigma_T = 2\pi \int_0^\pi \frac{d\sigma_T}{d\Omega} \sin\theta d\theta \tag{12.2}$$

Then, $8\pi r_e^2/3$ is obtained.

Rayleigh scattering results when the electron is bound with an atom and not a free electron. Because all atomic electrons behave similarly, the radiation scattered by the individual bound electrons of a given atom will be coherent—i.e., capable of showing constructive or destructive interference. This type of scattering is also called coherent scattering. Similar to Thomson scattering, the deflection angle is very small and energy loss is nothing. The differential cross section is represented using the Thomson cross section,

$$\frac{d\sigma_{coh}}{d\Omega} = F^2(Z, v)\frac{d\sigma_T}{d\Omega} \tag{12.3}$$

where Z is the atomic number and v is a parameter that represents the momentum transfer to an atom. If photon energy $h\nu$ is in a keV unit, v is written by in angstrom unit:

$$v = \frac{h\nu}{12.4} \sin \frac{\theta}{2} \quad \mathring{A}^{-1} \tag{12.4}$$

EXAMPLE 12.1

Calculate the parameter v for the momentum transfer for the incident photon of energy 1 MeV scattered through the angle 45° relative to the beam's direction.

SOLUTION 12.1

From Equation (12.4), v is calculated from

$$v = \frac{h\nu \,(\text{keV})}{12.4} \sin \frac{\theta}{2} \quad \mathring{A}^{-1} \quad = \frac{1000}{12.4} \sin \frac{45}{2} \quad \mathring{A}^{-1} \quad = 30.86 \quad \mathring{A}^{-1}$$

The atomic form factor $F(Z,v)$ represents the probability that an orbital electron receives the recoil momentum v. This factor has an effect decreasing the Thomson cross section at backward scattering. Hubbell et al. (1975) provides the table of $F(Z,v)$ as a function of v. Figure 12.1 shows angular

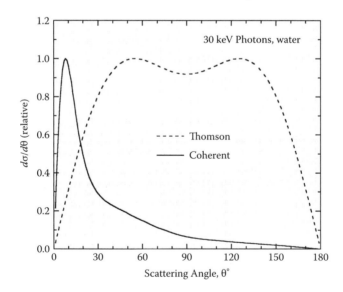

FIGURE 12.1
Comparison of angular distributions between coherent scattering and Thomson scattering.

distribution $d\sigma/d\theta$ for coherent scattering in comparison with Thomson scattering in which $d\sigma/d\theta$ (relative) was obtained by multiplying $2\pi\sin\theta$ by Equation (12.1).

An example on Rayleigh scattering is seen near ourselves. In the visible light region with energies lower than x-rays, the cross section for Rayleigh scattering is represented by

$$\frac{d\sigma_{\text{coh}}}{d\Omega} = \left(\frac{2\pi}{\lambda}\right)^4 |\alpha(\lambda)|^2 \frac{1}{2}(1+\cos^2\theta) \tag{12.5}$$

where λ is the wavelength of photons, and $\alpha(\lambda)$ is a quantity dependent on wavelength. It is noticed that angular distribution has a dependence of $1/\lambda^4$. When we observe the scattering at a θ direction, the cross section for a blue light with a short wavelength becomes much larger than that for a red light with a long wavelength. Therefore, a blue light significantly deflects. In the scattering of sunshine by atmospheric molecules, blue light is strongly scattered. This is the reason for the blue sky.

12.2 Photoelectric Effect

The theoretical calculation on photoelectric cross sections is very complicated. Here just the tendency is noted in place of the explicit formula. The photoelectric cross section depends on the atomic number Z and the photon energy $h\nu$. That is enhanced for matter with the large Z and for the low-energy photons with the frequency exceeding the threshold ν_0. For energies lower than the K absorption edge, the photoelectric absorption cross section is in proportion to Z^5 and is in inverse proportion to $(h\nu)^{7/2}$. In the region around 500 keV, it is in inverse proportion to $(h\nu)^2$. If $h\nu \gg mc^2$, it is in inverse proportion to $h\nu$. It is experimentally confirmed that the cross section for the other orbital electron is about 20% of that for the K-electron. Angular distribution of the photoelectrons is approximated in the nonrelativistic region as

$$\frac{d\sigma}{d\Omega} \propto \frac{\sin^2\theta}{(1-\beta\cos\theta)^4} \tag{12.6}$$

The probability of forward emission increases for $\beta \sim 1$. Figure 12.2 shows the angular distributions of photoelectrons with various β.

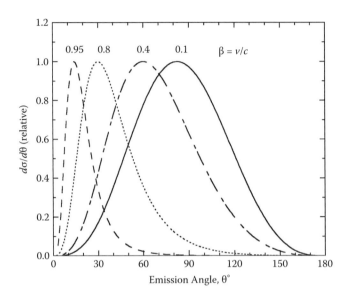

FIGURE 12.2
Emission angles of photoelectrons for various β values, in which β is the ratio of the initial speed of photoelectrons to the light speed.

12.3 Incoherent Scattering

The angular distribution of a scattered photon is given by Klein and Nishina.

$$\frac{d\sigma_{KN}}{d\Omega} = \frac{r_e^2}{2}\left(\frac{\nu'}{\nu}\right)^2\left(\frac{\nu}{\nu'} + \frac{\nu'}{\nu} - \sin^2\theta\right) \tag{12.7}$$

This formula is written inserting Equation (7.4) as

$$\frac{d\sigma_{KN}}{d\Omega} = \frac{r_e^2}{2}\left\{\frac{1}{1+\alpha(1-\cos\theta)}\right\}^2\left[1+\cos^2\theta + \frac{\alpha^2(1-\cos\theta)^2}{1+\alpha(1-\cos\theta)}\right] \tag{12.8}$$

in which $\alpha = h\nu/mc^2$. For $\alpha \ll 1$ this results in the Thomson cross section (Equation (12.1)). Taking into account the atomic structure of matter, scattering by many electrons is incoherent. Therefore, Compton scattering is incoherent scattering. The cross section of the incoherent scattering is given by

$$\frac{d\sigma_{incoh}}{d\Omega} = S(Z,\nu)\frac{d\sigma_{KN}}{d\Omega} \tag{12.9}$$

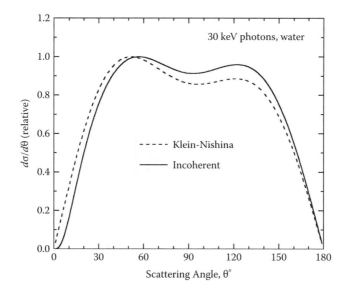

FIGURE 12.3
Comparison of angular distributions between incoherent scattering and Compton scattering.

Scattering function $S(Z,v)$ is the probability that an orbital electron received the photon momentum and excites or ionizes the atom. This effect reduces the forward angle scattering in the Klein-Nishina formula. Figure 12.3 shows a comparison of angular distributions between incoherent scattering and Compton scattering for 30 keV photons in water. The energy spectrum for the recoiled electron becomes

$$\frac{d\sigma_{KN}(T)}{dT} = \frac{\pi r_e^2}{\alpha^2(mc^2)} \left\{ 2 + \left[\frac{T}{h\nu - T} \right]^2 \left[\frac{1}{\alpha^2} + \frac{h\nu - T}{h\nu} - \frac{2}{\alpha} \frac{h\nu - T}{T} \right] \right\} \quad (12.10)$$

Figure 12.4 shows the energy spectrum of the first generation of recoiled electrons generated in copper irradiated with ^{137}Cs photons with the energy of 662 keV. The line spectrum at 653 keV is the photoelectron emitted from the K-shell of Cu.

EXAMPLE 12.2

Compare the intensity of photons scattered by coherent and incoherent scatterings through the angle 45° relative to the incident photon beam that impinges a tungsten target. The coherent scattering factors $F(Z = 74, v)$ for $v = 20$ and 50 Å^{-1} are given by 0.75777 and 0.10849, respectively, and the incoherent scattering factors $S(Z = 74, v)$ for $v = 20$ and 50 Å^{-1} are 73.711 and 73.998, respectively (Hubbell et al. 1975).

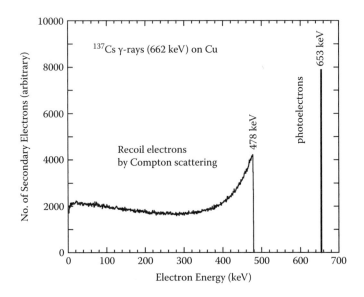

FIGURE 12.4
Energy spectra of the first generation of secondary electrons generated in copper irradiated with ^{137}Cs γ-rays with the energy of 662 keV.

Solution 12.2

From Example 12.1, $v = 30.86\,\text{Å}^{-1}$, $F(Z = 74, v = 30.86)$, and $S(Z = 74, v = 30.86)$ can be interpolated, for instance, using the logarithmic scale, from the tabulated data, i.e.,

$F(Z = 74, v = 30.86)$

$$= \exp\left[\frac{\ln(F(74, 50)) - \ln(F(74, 20))}{\ln(50) - \ln(20)} \cdot (\ln(30.86) - \ln(20)) + \ln(F(74, 20))\right]$$

$$= \exp\left[\frac{\ln(0.10849) - \ln(0.75777)}{\ln(50) - \ln(20)} \cdot (\ln(30.86) - \ln(20)) + \ln(0.10849)\right]$$

$$= 0.301966$$

In the same manner, $S(Z = 74, v = 30.86) = 73.84671$.
For coherent scattering, the differential cross section is calculated from:

$$\frac{d\sigma_{coh}}{d\Omega} = F^2(Z, v)\frac{d\sigma_T}{d\Omega}$$

$$= (0.301966)^2 \cdot \frac{(2.81794 \cdot 10^{-13} \text{ cm })^2}{2}(1 + \cos^2 45°)$$

$$= 5.43 \cdot 10^{-27} \text{ cm}^2/\text{sr}$$

The incoherent scattering cross section is given by

$$\frac{d\sigma_{incoh}}{d\Omega} = S(Z,v) \cdot \frac{r_e^2}{2} \left\{ \frac{1}{1+\alpha(1-\cos\theta)} \right\}^2 \left[1+\cos^2\theta + \frac{\alpha^2(1-\cos\theta)^2}{1+\alpha(1-\cos\theta)} \right]$$

$$= 73.84671 \cdot \frac{(2.81794 \cdot 10^{-13} \text{ cm})^2}{2} \left\{ \frac{1}{1+\frac{1}{0.511}(1-\cos 45°)} \right\}^2$$

$$\times \left[1+\cos^2 45 + \frac{\left(\frac{1}{0.511}\right)^2 (1-\cos 45°)^2}{1+\left(\frac{1}{0.511}\right)(1-\cos 45°)} \right]$$

$$= 2.62 \cdot 10^{-24} \text{ cm}^2/\text{sr}$$

The intensity of the photons scattered through the angle 45° is therefore 483 times larger by incoherent scatterings than those scattered by coherent (elastic) scatterings.

EXAMPLE 12.3
Compare the frequencies for detection of Compton electrons of energies 10, 100, and 800 keV, using the incident photon beam of energy 1 MeV.

SOLUTION 12.3
From Equation (6.6), the maximal energy of Compton electrons is

$$T_{max} = \frac{2hv}{2+mc^2/hv} = \frac{2 \cdot 1 \text{ MeV}}{2+0.511/1} = 0.786 \text{ MeV}$$

Therefore, it is not possible to detect 800 keV Compton electrons.
The frequency of detecting 10 keV Compton electrons is associated with the electron spectrum, as follows:

$$\frac{d\sigma_{KN}(0.01 \text{ MeV})}{dT}$$

$$= \frac{\pi r_e^2}{\alpha^2(mc^2)} \left\{ 2 + \left[\frac{T}{hv-T} \right]^2 \left[\frac{1}{\alpha^2} + \frac{hv-T}{hv} - \frac{2}{\alpha}\frac{hv-T}{T} \right] \right\}$$

$$= \frac{\pi \cdot (2.81794 \cdot 10^{-13} \text{ cm })^2}{\left(\frac{1}{0.511}\right)^2 (0.511 \text{ MeV})}$$

$$\times \left\{ 2 + \left[\frac{0.01}{1-0.01} \right]^2 \left[\frac{1}{\left(\frac{1}{0.511}\right)^2} + \frac{1-0.01}{1} - \frac{2}{\left(\frac{1}{0.511}\right)}\frac{1-0.01}{0.01} \right] \right\}$$

$$- 6.623 \cdot 10^{-26} \text{ cm}^2/\text{MeV}$$

For 100 keV Compton electrons, the corresponding cross section is $6.310 \cdot 10^{-26}$ cm^2/MeV. The frequency for detecting the smaller-energy Compton electrons (10 keV) is therefore only 5% larger than that for higher-energy electrons (100 keV). (You may compare the similarity of this result with Figure 12.4.)

12.4 Pair Creation

The cross section of pair creation increases as the photon energy increases and in proportion to Z^2. The cross section formula for positron production is given by Hough (1948).

$$\frac{d\sigma_{PC}}{dT_+} = \sigma_0 Z[1 + 0.135(\sigma_0 - 0.52)Z(1 - Z^2)] \qquad (12.11)$$

where

$$\sigma_0 = (1 - \gamma)[(4 - \gamma^2)(L - 1)/3 - \gamma^2\alpha(\alpha - 1) - \gamma^4 \alpha(L - \alpha)] \text{ and } Z = 2\sqrt{x(1 - x)} \qquad (12.12a)$$

in which

$$k = h\nu/mc^2, \quad E_+ = T_+ + mc^2, \quad x = (E_+ - 1)/(k - 2), \quad \gamma = 2/k,$$

$$L = \frac{2}{1 - \gamma^2}\ln\frac{k}{2}, \quad \alpha = \frac{1}{\sqrt{1 - \gamma^2}} \quad \ln\left[\frac{k}{2} + \sqrt{\left(\frac{k}{2}\right)^2 - 1}\right] \qquad (12.12b)$$

Figure 12.5 shows the energy spectra of positrons produced by photons of 10, 30, and 50 MeV.

The total cross sections are approximately given by

$$\sigma \sim Z^2(h\nu - 2mc^2), \quad (h\nu \geq 2mc^2)$$

$$\sigma \sim Z^2 \log h\nu, \quad (h\nu \gg 2mc^2) \qquad (12.13)$$

The average emission angle for electrons and positrons for the photon direction is approximated by

$$\theta_\pm \approx mc^2/T_\pm \qquad (12.14)$$

and an azimuth angle between electrons and positrons is 180°.

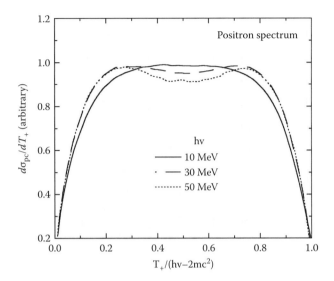

FIGURE 12.5
Energy spectra of positrons produced by pair creation with the photon energies 10, 30, and 50 MeV.

There are a lot of published data for photon interactions with matter in which the cross sections are provided in barn/atom units, and coherent and incoherent cross sections are given taking into account the atomic structure. Figures 12.6 and 12.7 show the total cross sections of photon interactions with water and lead, respectively, tabulated by Storm and Israel (1970).

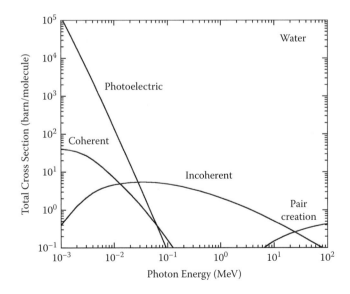

FIGURE 12.6
Total cross sections of photon interactions with water.

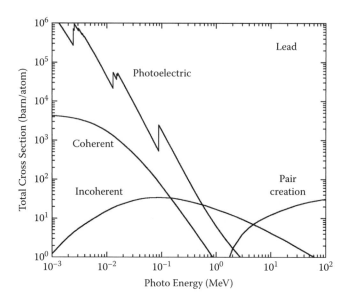

FIGURE 12.7
Total cross sections of photon interactions with lead.

EXAMPLE 12.4

Calculate the differential cross sections for production of 500 keV positrons by a 10 MeV photon beam hitting a tungsten target. Find the average angles that these positrons are scattered through. Compare the intensity for detecting these positrons with the intensity of 5 MeV positrons, given that the collision system is the same.

SOLUTION 12.4

The number of detected positrons is calculated from the energy spectrum of the positron, given by Equation (12.11):

$$\frac{d\sigma_{PC}}{dT_+} = \sigma_0 z[1 + 0.135(\sigma_0 - 0.52)\, z(1 - z^2)]\,.$$

The cross sections unit is

$$\frac{Z^2 a_0^2}{137}\Big/ mc^2\,,$$

and

$$k = h\nu/mc^2 = 10/0.511 = 19.569$$

$$\gamma = 2/k = 2/19.569 = 0.1022$$

$$\alpha = \frac{1}{\sqrt{1-\gamma^2}} \ln\left[\frac{k}{2} + \sqrt{\left(\frac{k}{2}\right)^2 - 1}\right]$$

$$= \frac{1}{\sqrt{1-0.1022^2}} \ln\left[\frac{19.569}{2} + \sqrt{\left(\frac{19.569}{2}\right)^2 - 1}\right]$$

$$= 2.987$$

$$L = \frac{2}{1-\gamma^2} \ln\frac{k}{2} = \frac{2}{1-0.1022^2} \ln\frac{19.569}{2} = 4.610$$

$$\sigma_0 = (1-\gamma)[(4-\gamma^2)(L-1)/3 - \gamma^2\alpha(\alpha-1) - \gamma^4\alpha(L-\alpha)]$$

$$= (1-0.1022)[(4-0.1022^2)(4.610-1)/3 - 0.1022^2 \cdot 2.987(2.987-1)$$

$$- 0.1022^4 \cdot 2.987(4.610-2.987)]$$

$$= 4.254$$

$$E_+ = T_+ + mc^2 = 0.5/0.511 + 1 = 1.978$$

$$x = (E_+ - 1)/(k-2) = (1.978-1)/(19.569-2) = 0.0557$$

$$z = 2\sqrt{x(1-x)} = 2\sqrt{0.0557(1-0.0557)} = 0.459$$

Finally,

$$\frac{d\sigma_{PC}(500\ \text{keV})}{dT_+} = \sigma_0 \cdot 0.459 \cdot [1 + 0.135 \cdot (4.254-0.52) \cdot 0.459 \cdot (1-0.459^2)]$$

$$= 4.254 \cdot 0.5424$$

$$= 2.307 \cdot \frac{Z^2 a_0^2}{137} \Big/ mc^2$$

$$= \frac{2.307 \cdot 74^2 \cdot (5.29 \cdot 10^{-9}\ \text{cm})^2}{137 \cdot 511\ \text{keV}} = 5.05 \cdot 10^{-18}\ \text{cm}^2/\text{keV}$$

The average angle is $\theta_\pm \approx 0.511/0.5 = 1.022\ \text{sr} = 58.56°$.
   For 5 MeV positrons:

$$E_+ = T_+ + mc^2 = 5/0.511 + 1 = 10.785$$

$$x = (E_+ - 1)/(k-2) = (10.785-1)/(19.569-2) = 0.557$$

$$z = 2\sqrt{x(1-x)} = 2\sqrt{0.557(1-0.557)} = 0.993$$

$$\frac{d\sigma_{PC}}{dT_+} \approx \sigma_0$$

It is to note that 5 MeV is about half of the energy transferred to the electron-positron pair, for which the energy spectrum exhibits a maximum (cf. Figure 12.5).

Therefore, the intensity ratio is

$$\frac{d\sigma_{PC}(500 \text{ keV})}{dT_+} \Big/ \frac{d\sigma_{PC}(100 \text{ keV})}{dT_+} = 0.5424/1 = 0.5424$$

12.5 Soft X-Rays

Koch and Motz (1959) summarized the theory of bremsstrahlung. Symbols used here are as follows, in which the unit of energy is in electron rest mass (mc^2).

Z = atomic number of target material

T = initial kinetic energy of the electron in a collision

E = initial total energy of the electron in a collision ($= T + 1$)

k = energy of the emitted photon

E' = final total energy of the electron in a collision ($= E - k$)

γ = screening parameter ($= 100k/EE'Z^{1/3}$); complete screening ($\gamma = 0$), nonscreened case ($\gamma = \infty$)

The differential cross section of bremsstrahlung for low-energy electrons in the range $10 \text{ keV} \leq T < 2 \text{ MeV}$ is given by

$$\frac{d\sigma}{dk} = A_E f_E \frac{d\sigma^{3BN}}{dk} \tag{12.15}$$

where f_E is called the Elwert correction factor,

$$f_E = \frac{\beta_0 \left[1 - e^{-2\pi Z/137\beta_0}\right]}{\beta \left[1 - e^{-2\pi Z/137\beta}\right]} \tag{12.16}$$

$$\beta_0^2 = 1 - 1/E^2, \quad \beta^2 = 1 - 1/E'^2$$

and the correction factor A_E is determined to fit to the experimental data. The cross section $d\sigma^{3BN}/dk$ is obtained using the Born approximation under the nonscreened case.

$$\frac{d\sigma^{3BN}}{dk} = \frac{Z(Z+1)r_e^2}{137} \frac{p'}{p} \left[\frac{4}{3} - 2EE' \left(\frac{p'^2 + p^2}{p'^2 p^2} \right) + \frac{\varepsilon E'}{p^3} + \frac{\varepsilon' E}{p'^3} - \frac{\varepsilon \varepsilon'}{pp'} + LU \right] \frac{1}{k} \tag{12.17}$$

where

$$r_e = 2.81794 \times 10^{-13} \quad \text{cm}, \quad p^2 = E^2 - 1, \quad p'^2 = E'^2 - 1$$

$$\varepsilon = \ln \frac{E + p}{E - p}, \quad \varepsilon' = \ln \frac{E' + p'}{E' - p'}$$

$$L = 2 \ln \left(\frac{EE' + pp' - 1}{k} \right)$$

$$U = \frac{8EE'}{3pp'} + k^2 \frac{E^2 E'^2 + p^2 p'^2}{p^3 p'^3} + \frac{k}{2pp'} \left[\left(\frac{EE' + p^2}{p^3} \right) \varepsilon - \left(\frac{EE' + p^2}{p'^3} \right) \varepsilon' + \frac{2kEE'}{p^2 p'^2} \right]$$

$$(12.18)$$

Figure 12.8 shows the bremsstrahlung spectra produced by electrons with the energies 50, 100, and 500 keV incident on thin tungsten target.

The Koch-Motz formula for high-energy electrons in the range 2 MeV $\leq T <$ 50 MeV is classified according to both the electron energy and the screening parameter γ.

2 MeV $\leq T <$ 15 MeV	$d\sigma/dk = A_E \, d\sigma^{3BN}/dk$	$\gamma > 15$	(12.19a)
	$d\sigma/dk = A_E \, d\sigma^{3BSd}/dk$	$2 \leq \gamma \leq 15$	(12.19b)
	$d\sigma/dk = A_E \, d\sigma^{3BSc}/dk$	$\gamma < 2$	(12.19c)

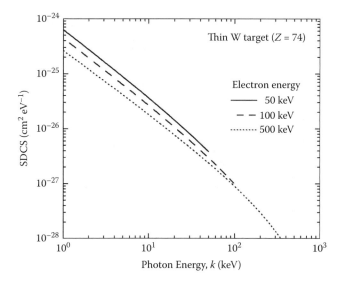

FIGURE 12.8

Bremsstrahlung spectra produced by electrons with the energies 50, 100, and 500 keV incident on thin tungsten target.

and

$$15 \text{ MeV} \le T < 50 \text{ MeV} \qquad d\sigma/dk = d\sigma^{3BN}/dk \qquad \gamma > 15 \qquad (12.20a)$$

$$d\sigma/dk = A_E\, d\sigma^{3BSd}/dk \qquad 2 \le \gamma \le 15 \qquad (12.20b)$$

$$d\sigma/dk = A_E\, d\sigma^{3BSc}/dk \qquad \gamma < 2 \qquad (12.20c)$$

In Equations (12.19) and (12.20), the explicit forms for $d\sigma^{3BSd}/dk$ and $d\sigma^{3BSc}/dk$ are given as follows:

$$\frac{d\sigma^{3BSd}}{dk} = \frac{4Z(Z+1)r_e^2}{137}\left[1+\left(\frac{E'}{E}\right)^2 - \frac{2E'}{3E}\right]\left[\ln\frac{2EE'}{k} - \frac{1}{2} - c(\gamma)\right]\frac{1}{k}$$

$$c(\gamma) = 0.102e^{-0.151\gamma} + 0.47e^{-0.63\gamma} \qquad (12.21)$$

$$\frac{d\sigma^{3BSc}}{dk} = \frac{4Z(Z+1)r_e^2}{137}\left\{\left[1+\left(\frac{E'}{E}\right)^2\right]\left[\frac{\phi_1(\gamma)}{4} - \frac{\ln Z}{3}\right] - \frac{2E'}{3E}\left[\frac{\phi_2(\gamma)}{4} - \frac{\ln Z}{3}\right]\right\}\frac{1}{k}$$

$$\phi_1(\gamma) = \phi_2(\gamma) + 0.5e^{-2.31\gamma} + 0.12e^{-19.8\gamma}, \quad \phi_2(\gamma) = 20.14e^{-0.151\gamma}$$

$$(12.22)$$

We consider the realistic spectrum such as the accumulated spectrum of all photons generated by electrons until they completely stop in a sufficiently thick medium. This spectrum is called a thick target. It is assumed the number of photons within the energy interval $(k, k + dk)$ produced by an electron of the kinetic energy T is dn, and the number of the ith atom contained in 1 g of material is N_i. The thick target is regarded as many layers of thin targets piled up. Taking into account the energy loss by the layers, electrons gradually slow down. Therefore, the spectrum changes layer by layer. The aimed spectrum is obtained by accumulating these spectra. The number of photons can be calculated by the equation

$$dn = \sum_i N_i \int_{k+mc^2}^{T+mc^2} \frac{\frac{d\sigma_i}{dk}dk}{\left(-\frac{dE}{\rho dx}\right)_{\text{tot}}} dE \qquad (12.23)$$

The total mass stopping power at the denominator is the sum of the collision stopping power and the radiative stopping power. Figure 12.9 shows the calculated spectrum for thick tungsten target, in which absorption of low-energy photons is not considered.

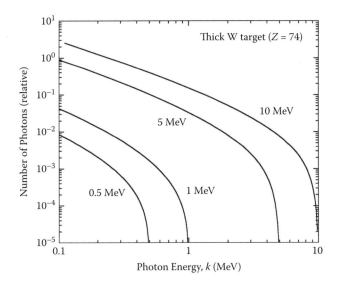

FIGURE 12.9
Bremsstrahlung spectra produced by electrons with the energies 0.5, 1, 5, and 10 MeV incident on thick tungsten target.

Differential cross sections in photon energy and angle were calculated using Schiff's formula:

$$\frac{d^2\sigma}{dkd\Omega}$$

$$= \frac{4Z(Z+1)r_e^2}{137} \frac{y}{k} \left\{ \frac{16y^2E'}{(y^2+1)^4E} - \frac{(E+E')^2}{(y^2+1)^2E^2} + \left[\frac{E^2+E'^2}{(y^2+1)^2E^2} - \frac{4y^2E'}{(y^2+1)^4E} \right] \ln M(y) \right\}$$

$$y = E\theta, \quad \frac{1}{M(y)} = \left(\frac{k}{2EE'} \right)^2 + \left(\frac{Z^{1/3}}{111(y^2+1)} \right)^2$$

$$(12.24)$$

Figure 12.10 shows the angular distributions for bremsstrahlung produced by electrons with the energies 50, 500 keV and 5 MeV incident on tungsten target. The relative angular distributions normalized at the maximum do not change so seriously between various photon energies for any electron energies.

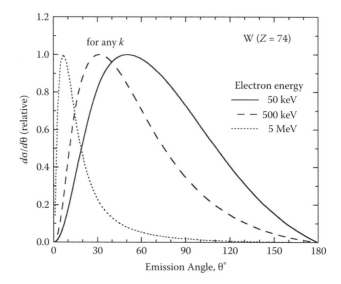

FIGURE 12.10
Angular distributions for bremsstrahlung photons produced by various electron energies. The values of k are not specified because the curves are almost similar for any photon energies.

12.6 Summary

1. Thomson scattering is modified by the atomic form factor taking into account electrons bound with atom. This is called coherent scattering.

2. Compton scattering is modified by the scattering function taking into account electrons bound with atom. This is called incoherent scattering.

3. Photoelectric cross sections are enhanced for high Z matter and for low-energy photons exceeding the absorption edge.

4. Cross sections of pair creation increase as the photon energy increases and in proportion to Z^2.

5. Bremsstrahlung spectra for thin and thick targets can be evaluated using the cross section formula given by Koch and Motz.

QUESTIONS

1. a. What are the K x-rays energies for carbon, aluminum, and titanium?

   b. What are the most probable electron energies emitted from interaction of C_K, Al_K, and Ti_K x-rays in water?

2. A source of C (Ci) of some radionuclide emits *f* monoenergetic γ-rays of energy *E* (MeV) for every nuclear disintegration. At a distance *L* (m) from the source, we placed a volume of air 0.01 (m) thick and 0.01 (cm²) cross section. Neglecting energy loss in air in the *L* path, calculate:
   a. The flux *F* through the small volume
   b. Energy absorption by the volume
   c. The exposure rate
   d. The exposure rate, if the source emits several l-rays, each with characteristic energy, fractional occurrence, and absorption coefficient

3. Write an expression for the mean free path length of a photon traveling through a homogenous medium.

4. Write down the formula of the Klein-Nishina differential cross section for inelastic photon scattering by a free electron. Design a flowchart for the Monte Carlo sampling of the probability distribution of a scattered photon.

5. How many photoelectrons are generated when an infinite water target is irradiated with 50 eV photons?

References

Hough PVC. 1948. Low energy pair production. *Phys. Rev.* 73: 266–267.

Hubbell JH, Veigele Wm J, Briggs EA, Brown RT, Cromer DT, Howerton RJ. 1975. Atomic form factors, incoherent scattering functions, and photon scattering cross sections. *J. Phys. Chem. Ref. Data* 4: 471–616.

Koch HW, Motz JW. 1959. Bremsstrahlung cross-section formulas and related data. *Rev. Mod. Phys.* 31: 920–955.

Storm E, Israel HI. 1970. Photon cross sections from 1 keV to 100 MeV for elements Z=1 to Z=100. *Nucl Data Tables* A7: 565–681.

13

Cross Sections for Interactions of Electrons with Water

13.1 Ionization

As energetic electrons pass through matter they lose energy primarily through collisions with bound electrons. Ionization cross sections for all primary and secondary electrons are needed to follow the history of an incident particle and its products, covering all ranges of energy transferred in individual collisions. Cross section data for liquid water are scarce, as measurements are either impractical or very difficult (see Section III). In this section we describe the KURBUC code system, which uses water vapor cross sections, differential, total, and partial cross sections for ionization.

13.1.1 Secondary Electrons

The calculation of the secondary electron spectrum was carried out using the method of Seltzer (1988). For the jth orbital of a molecule, the cross section differential in kinetic energy ε of the ejected electron is written as the sum of contributions by the close collision and the distant collision.

$$\frac{d\sigma^{(j)}}{d\varepsilon} = \frac{d\sigma_c^{(j)}}{d\varepsilon} + \frac{d\sigma_d^{(j)}}{d\varepsilon} \tag{13.1}$$

The first term is described in terms of a collision between two electrons.

$$\frac{d\sigma_c^{(j)}}{d\varepsilon} = \frac{2\pi r_e^2 m_e c^2 n_j}{\beta^2} \frac{T}{T + B_j + U_j}$$

$$\times \left\{ \frac{1}{E^2} + \frac{1}{(T-\varepsilon)^2} + \frac{1}{T^2}\left(\frac{\tau}{\tau+1}\right)^2 - \frac{1}{E(T-\varepsilon)}\frac{2\tau+1}{(\tau+1)^2} + G_j \right\} \tag{13.2}$$

$$G_j = \frac{8U_j}{3\pi}\left[\frac{1}{F_i^3} + \frac{1}{(T-\varepsilon)^3}\right]\left[\tan^{-1}\sqrt{y} + \frac{\sqrt{y}(y-1)}{(y+1)^2}\right] \tag{13.3}$$

173

in which $r_e = 2.81794 \times 10^{-13}$ cm is the classical electron radius, $m_e c^2 = 511003.4$ eV is the electron rest mass, $\tau = T/m_e c^2$ is the kinetic energy in units of the rest mass, $\beta^2 = 1 - 1/(\tau + 1)^2$, n_j is the number of electrons in the orbital, B_j is the orbital binding energy, U_j is the mean kinetic energy of the target electron in the orbital, E is the energy transfer ($= \varepsilon + B_j$), and $y = \varepsilon/U_j$.

The second term of Equation (13.1) is described in terms of the interaction of the equivalent radiation field with the orbital electrons.

$$\frac{d\sigma_d^{(j)}}{d\varepsilon} = n_j I(E) \sigma_{PE}^{(j)}(E) \tag{13.4}$$

where $\sigma_{PE}^{(j)}$ is the photoelectric cross section for the jth orbital (per orbital electron), for an incident photon of energy $E = \varepsilon + B_j$. The virtual photon spectrum integrated over impact parameters $b_{min} < b < b_{max}$ is given by

$$I(E) = \frac{2}{137\pi\beta^2 E}[G(x_{min}) - H(x_{max})]$$

$$G(x_{min}) = x_{min}K_0(x_{min})K_1(x_{min}) - \frac{x_{min}^2}{2}\left\{K_1^2(x_{min}) - K_0^2(x_{min})\right\} \tag{13.5}$$

$$H(x_{max}) = x_{max}K_0(x_{max})K_1(x_{max}) - \frac{x_{max}^2}{2}\left\{K_1^2(x_{max}) - K_0^2(x_{max})\right\}$$

where

$$x_{min} = \frac{Eb_{min}}{\hbar c}\frac{\sqrt{1-\beta^2}}{\beta} \tag{13.6}$$

$$x_{max} = \frac{Eb_{max}}{\hbar c}\frac{\sqrt{1-\beta^2}}{\beta} \tag{13.7}$$

b_{min} and b_{max} are impact parameters. $b_{min} = <r>_j$ is the expectation value of the electron radius for the orbital of interest, and b_{max} is given by

$$b_{max} = \frac{1.123\hbar c\beta}{B_j\sqrt{1-\beta^2}} \tag{13.8}$$

K_0 and K_1 are the Bessel functions of order 0 and 1. It should also be noted that $I(E)$, x_{min}, and b_{min} are different for each orbital.

For numerical calculations for a water molecule, various input data were required. Input data for water vapor were used because they are more readily available than those for liquid water. Atomic data for each shell of hydrogen and oxygen are given in Table 13.1. The atomic electron radii $<r>$ listed in Table 13.1 were taken from the calculation of Mann (1968).

TABLE 13.1

Electron Numbers N and Mean Radius $<r>$ in Bohr Units a_B for Four Atomic Orbitals Composing the Molecular Orbitals of Water

Orbital	N	$<r>$
H_{1s}	1	1.0
O_{2p}	4	0.833
O_{2s}	2	0.875
O_{1s}	2	0.129

Atomic subshell photoelectric cross sections are shown in Figure 13.1. Data for photon energies $E < 8$ keV were taken from the evaluation by Yeh and Lindau (1985), and those for energies higher than 8 keV were extrapolated so as to satisfy the total photoelectric cross sections given by Hubbell (1977). The photoelectric cross sections for molecular orbitals were assembled from a linear combination of cross sections for atomic orbitals using the parentage coefficients given by Siegbahn (1969) as listed in Table 13.2. Values of B_j were taken from Zaider et al. (1983). Kinetic energies U_j of the orbital electron are values assumed by analogy with

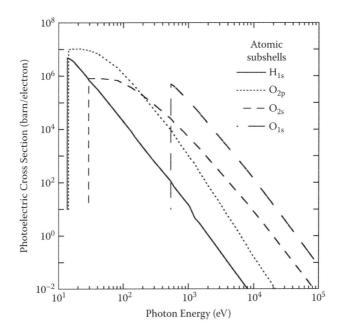

FIGURE 13.1

Atomic subshell photoelectric cross sections as a function of photon energy in the range of 10 eV–100 keV for hydrogen H_{1s} and oxygen O_{2p}, O_{2s}, and O_{1s} states

TABLE 13.2

Electron Number n_j, Binding Energy B_j, Kinetic Energy
U_j, Parentage Coefficients, and Mean Radius $<r>_j$ for the
jth Orbital of a Water Molecule

j	Orbital	n_j	B_j (eV)	U_j (eV)	Parentage Coefficients	$<r>_j$
1	$1b_1$	2	12.62	30	1.00 O_{2p}	0.833
2	$3a_1$	2	14.75	40	0.175 H_{1s}	0.867
					0.72 O_{2p}	
					0.105 O_{2s}	
3	$1b_2$	2	18.51	50	0.41 H_{1s}	0.901
					0.59 O_{2p}	
4	$2a_1$	2	32.4	60	0.25 H_{1s}	0.906
					0.75 O_{2s}	
5	O_{1s}	2	539.7	700	1.00 O_{1s}	0.129

N_2 data given by Kim (1975). Figure 13.2 presents the assembled $\sigma_{PE}^{(j)}$ per electron in the jth orbital of the molecule. The molecular radii $<r>_j$ were assembled using the atomic electron radii of Table 13.1 and the parentage coefficients of Table 13.2. Values of molecular radii $<r>_j$ were used for the impact parameter b_{min} in Equation (13.6). It was assumed that the

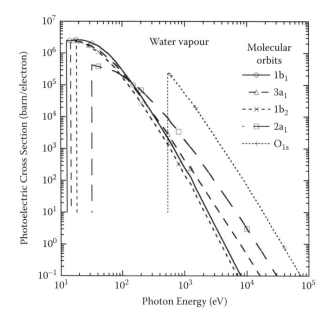

FIGURE 13.2
Molecular orbital photoelectric cross sections for water vapor for photon energies 10 eV–100 keV.

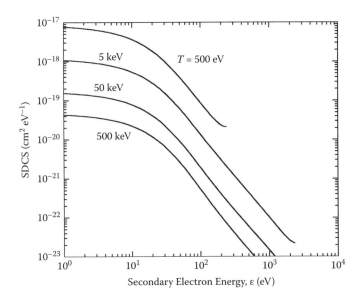

FIGURE 13.3
Energy spectra of secondary electrons for 0.5, 5, 50. and 500 keV electrons calculated using the Seltzer's theory.

Auger electrons were emitted isotropically from the inner shell, having an energy of 514.5 (= 539.7 − 12.62 × 2) eV. Coster-Kronig electrons were not treated.

Figure 13.3 shows the calculated single differential cross sections (SDCSs) of secondary electrons for primary electron energies, $T = 0.5$, 5, 50, and 500 keV. For $T = 1$ and 10 keV, a good agreement with the calculated data of Paretzke (1988) and the experimental data of Vroom and Palmer (1977) and Bolorizadeh and Rudd (1986) is verified (Nikjoo et al. 2006).

Angular distribution of secondary electrons for the ejected energy $\varepsilon <$ 200 eV, regardless of the primary electron energy, was randomly sampled using the experimental doubly differential cross sections of Opal et al. (1972). Figure 13.4 shows the sampled distributions for various ε (eV) ejected by electron impact with any values of T. For $\varepsilon \geq 200$ eV produced by any primary energy, the scattering angle was approximated using a kinematical relationship:

$$\cos\theta = \sqrt{\frac{\varepsilon\left(T + 2m_0c^2\right)}{T\left(\varepsilon + 2m_0c^2\right)}} \tag{13.9}$$

Thus scattering angle is uniquely determined as a function of T and ε. Figure 13.5 shows the relationship between three parameters, T, ε, and θ.

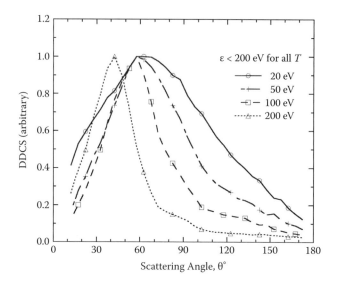

FIGURE 13.4
Angular distributions of secondary electrons with various energies smaller than 200 eV for all energies of the primary electron.

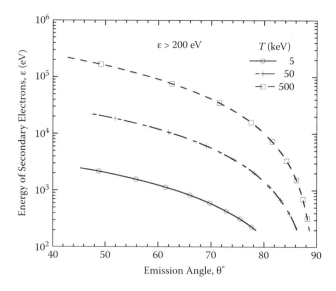

FIGURE 13.5
Kinematical relationship between the emission angle and the energy of secondary electrons greater than 200 eV for the primary electron energy, *T*.

EXAMPLE 13.1

Calculate the ejected electron energy of the scattering angle 70° for the incident electron energy 5 keV.

SOLUTION 13.1

$$\cos\theta = \sqrt{\frac{\varepsilon(T + 2m_0c^2)}{T(\varepsilon + 2m_0c^2)}}$$

$$\cos 70° = \sqrt{\frac{\varepsilon(5 + 1022)}{5(\varepsilon + 1022)}}$$

$$\varepsilon = 0.58 \text{ keV}$$

13.1.2 Total Cross Sections

Total ionization cross sections (TICSs) were compiled from different sources for low- and high-energy electrons (Uehara et al. 1999). In the low-energy region between 10 eV and 10 keV, all available experimental ionization cross sections were least squares fitted, including those of Hayashi (1989), Djuric et al. (1988), Olivero et al. (1972), Bolorizadeh and Rudd (1986), and Schutten et al. (1966). The fitted function is given by

$$\sigma_{\text{exp}} = c + at^r e^{-(b_1 t + b_2 t^2)}$$

(13.10)

$$t = \ln\frac{T}{15}$$

in which the unit of cross section is 10^{-16} cm^2 and that of T, the particle kinetic energy, is eV. The fitting parameters are given by

$$a = 2.07201 \qquad b_1 = 0.271302 \qquad b_2 = 0.119638$$
$$r = 1.46521 \qquad c = 0.074664$$

For the energies higher than 10 keV up to the MeV region, ionization cross sections were calculated using Seltzer's formula:

$$\sigma_{\text{ion}} = \sum_{j=1}^{5} \sigma^{(j)} = \sum_{j=1}^{5} \int_0^{(T-B_j)/2} \frac{d\sigma^{(j)}}{d\varepsilon} d\varepsilon$$

(13.11)

in which $\sigma^{(j)}$ is the partial ionization cross section for the jth molecular orbital. This $\sigma^{(j)}$ was applied to the low-energy region less than 10 keV. The fraction of partial ionization cross section for a subshell is given by

$$f_j = \sigma^{(j)}/\sigma_{\text{ion}}$$

(13.12)

Figure 13.6 shows the fractions for five molecular orbits of water. The TICS smoothly connected at the boundary of 10 keV is shown in Figure 13.7.

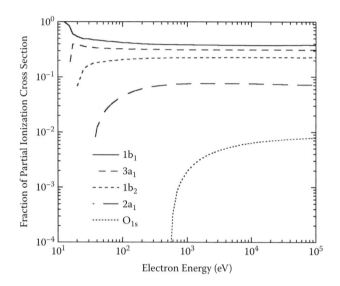

FIGURE 13.6
Fractions of partial ionization cross sections for the subshells $1b_1$, $3a_1$, $1b_2$, $2a_1$ and the oxygen O_{1s} inner shell state.

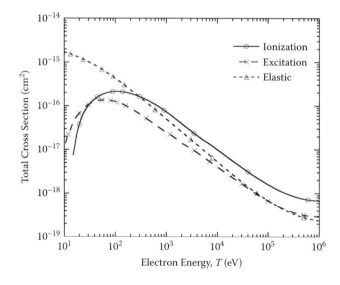

FIGURE 13.7
Total cross sections for ionization, excitation, and elastic scattering of electrons for water vapor.

EXAMPLE 13.2

Calculate the total cross section for an electron with 50 eV energy.

SOLUTION 13.2

$$t = \ln\frac{T}{15} = \ln\frac{50}{15} = 1.204$$

$$\sigma_{\exp} = c + at^r e^{-(b_1 t + b_2 t^2)}$$

$$= 0.074664 + 2.07201 \times 1.204^{1.46521} e^{-(0.271302 \times 1.204 + 0.119638 \times 1.204^2)}$$

$$= 1.72 \times 10^{-16}\,\mathrm{cm}^2$$

13.2 Excitation

There are various modes of excitation for $T > 7.4$ eV. As the reported experimental excitation cross sections are fragmentary, we used the compiled data of Paretzke (1988), which encompass almost all the major excitation modes. Paretzke fitted the experimental cross sections for the major individual states using the model function:

$$\sigma_{\exc} = \Phi_3(T)\frac{4\pi a_0^2 R}{T} M_a^2 \ln\frac{4c_s T}{R} \tag{13.13a}$$

in which $\Phi_3(T)$ is an empirical correction factor,

$$\Phi_3(T) = 1 - e^{-0.25(T/E_a - 1)} \tag{13.13b}$$

where a_0 is the Bohr radius, 0.529×10^{-8} cm, and R is the Rydberg constant equal to 13.6 eV. Table 13.3 lists the parameters M_a^2, E_a, and c_s for each level

TABLE 13.3

Parameters for Seven Major Excitation States
Taken from Paretzke (1988)

Excitation State	E_a (eV)	M_a^2	c_s
A^1B_1	7.4	0.099	1.25
B^1A_1	9.7	0.098	1.25
Diffuse band	13.3	0.363	1.25
Rydberg (A + B)	10.0	0.041	1.25
Rydberg (C + D)	11.0	0.072	1.25
H* Lyman a	21.0	0.088	115
H* Balmer a	21.0	0.0206	32

considered in KURBUC. An additional OH* level was fitted by the analytical model given by Green and Stolarski (1972).

$$\sigma_{exc} = \frac{4\pi a_0^2 R^2}{E_a^2} f_0 c_0 \left(\frac{E_a}{T}\right)^{\Omega} \left\{1-\left(\frac{E_a}{T}\right)^{\beta}\right\}^{\nu} \tag{13.14}$$

in which the parameters given by Paretzke (1988) are $E_a = 9.0$ eV, $f_0 c_0 = 0.033$, $\Omega = 1$, $\beta = 2$, and $\nu = 1$.

The total excitation cross sections in the energy range 10 eV to 10 keV were obtained by summing up the individual cross sections for the eight major levels listed above. The total excitation cross section for the high-energy region up to 10 MeV is given by an empirical formula derived from the so-called Fano plot, with the fitting parameters determined by Berger and Wang (1988):

$$\sigma_{exc} = 4\pi \left(\frac{a_0}{137\beta}\right)^2 \left[12.30 + 1.26 \left\{\ln \frac{\beta^2}{1-\beta^2} - \beta^2\right\}\right] \tag{13.15}$$

The total cross section is shown in Figure 13.7. Figure 13.8 shows the fractions of partial excitation cross sections for the major eight levels. The mean excitation energy $\overline{E_a}$ necessary to calculate the energy losses was derived according to the following equation:

$$\overline{E_a} = \sum_i E_a^i \sigma_{exc}^i \bigg/ \sum_i \sigma_{exc}^i \tag{13.16}$$

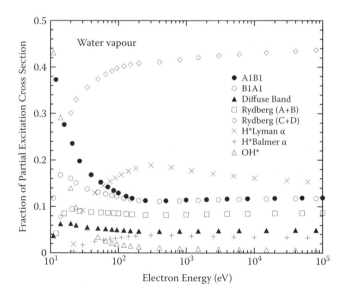

FIGURE 13.8
Fractions of partial excitation cross sections for eight major states.

in which $\sigma_{exc}{}^i$ and $E_a{}^i$ are the cross section and the excitation energy for the *i*th excitation mode, respectively. The values of $\overline{E}_a$ remain approximately constant around 13 eV in the 10 keV–10 MeV region.

EXAMPLE 13.3

Calculate the excitation cross section for 50 eV electrons.

SOLUTION 13.3

$$\sigma_{exc} = 1 - e^{-0.25(T/E_a - 1)} \frac{4\pi a_0^2 R}{T} M_a^2 \ln \frac{4c_s T}{R}$$

for A^1B_1

$$\sigma_{exc} = 1 - e^{-0.25(50/7.4 - 1)} \frac{4\pi(0.529 \times 10^{-8})^2 13.6}{50} 0.099 \times \ln \frac{4 \times 1.25 \times 50}{13.6}$$

$$= 1.55 \times 10^{-18}\,\text{cm}^2$$

for B^1A_1

$$\sigma_{exc} = 1 - e^{-0.25(50/9.7 - 1)} \frac{4\pi(0.529 \times 10^{-8})^2 13.6}{50} 0.098 \times \ln \frac{4 \times 1.25 \times 50}{13.6}$$

$$= 1.30 \times 10^{-18}\,\text{cm}^2$$

for Diffuse band

$$\sigma_{exc} = 1 - e^{-0.25(50/13.3 - 1)} \frac{4\pi(0.529 \times 10^{-8})^2 13.6}{50} 0.363 \times \ln \frac{4 \times 1.25 \times 50}{13.6}$$

$$= 3.70 \times 10^{-18}\,\text{cm}^2$$

for Rydberg $(A + B)$

$$\sigma_{exc} = 1 - e^{-0.25(50/10 - 1)} \frac{4\pi(0.529 \times 10^{-8})^2 13.6}{50} 0.041 \times \ln \frac{4 \times 1.25 \times 50}{13.6}$$

$$= 5.30 \times 10^{-19}\,\text{cm}^2$$

for Rydberg $(C + D)$

$$\sigma_{exc} = 1 - e^{-0.25(50/11 - 1)} \frac{4\pi(0.529 \times 10^{-8})^2 13.6}{50} 0.072 \times \ln \frac{4 \times 1.25 \times 50}{13.6}$$

$$= 8.66 \times 10^{-19}\,\text{cm}^2$$

for H^* Lyman α

$$\sigma_{exc} = 1 - e^{-0.25(50/21 - 1)} \frac{4\pi(0.529 \times 10^{-8})^2 13.6}{50} 0.088 \times \ln \frac{4 \times 115 \times 50}{13.6}$$

$$= 1.34 \times 10^{-18}\,\text{cm}^2$$

for H^* Balmer α

$$\sigma_{exc} = 1 - e^{-0.25(50/21 - 1)} \frac{4\pi(0.529 \times 10^{-8})^2 13.6}{50} 0.0206 \times \ln \frac{4 \times 32 \times 50}{13.6}$$

$$= 2.06 \times 10^{-19}\,\text{cm}^2$$

13.3 Elastic Scattering

Elastic scattering cross sections were calculated using the Rutherford formula taking into account the screening parameter η given by Moliere (1948). The differential cross section $(d\sigma/d\Omega)_{el}$ and total cross sections σ_{el} for each atom are represented by

$$\left(\frac{d\sigma}{d\Omega}\right)_{el} = Z(Z+1)r_e^2 \frac{1-\beta^2}{\beta^4} \frac{1}{(1-\cos\theta+2\eta)^2} \tag{13.17}$$

$$\sigma_{el} = \pi Z(Z+1)r_e^2 \frac{1-\beta^2}{\beta^4} \frac{1}{\eta(\eta+1)} \tag{13.18}$$

The screening parameter η is given by

$$\eta = \eta_c \times 1.7 \times 10^{-5} Z^{2/3} \frac{1}{\tau(\tau+2)} \tag{13.19}$$

where the effective atomic number of the water molecule, Z, was assumed to be 7.42, $r_e = 2.8179 \times 10^{-13}$ cm, $m_0c^2 = 511003.4$ eV is the electron rest mass, $\tau = T/m_0c^2$ is the kinetic energy in units of the rest mass, $\beta^2 = 1 - 1/(\tau + 1)^2$, and

$$\eta_c = 1.198 \qquad\qquad \text{for } T < 50 \text{ keV} \tag{13.20a}$$

$$= 1.13 + 3.76\left(\frac{Z}{137\beta}\right)^2 \quad \text{for } T \geq 50 \text{ keV} \tag{13.20b}$$

The parameter η_c was determined by the least squares fitting to the published experimental data (Shyn and Cho 1987; Katase et al. 1986; Nishimura 1979; Danjo and Nishimura 1985). Figure 13.7 shows the total cross sections for elastic scattering in the energy range 10 eV to 1 MeV.

Angular distributions for elastic scattering below 1 keV were obtained by direct sampling of various experimental data (Katase et al. 1986; Nishimura 1979; Trajmar et al. 1973) in place of Equation (13.17). Figure 13.9 shows the differential cross sections for electron elastic scattering in water ($d\sigma/d\Omega$ in relative units) as a function of scattering angle θ for various electron energies.

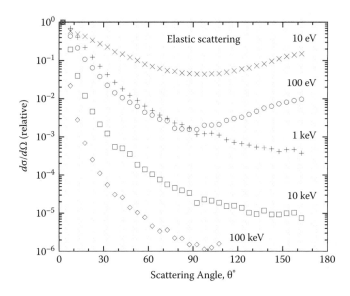

FIGURE 13.9
Differential cross sections for electron elastic scattering in water as a function of scattering angle θ for various electron energies.

EXAMPLE 13.4

Calculate the elastic scattering cross section for $T = 40$ keV.

SOLUTION 13.4

$$\eta_c = 1.198$$

$$\tau = \frac{T}{m_0 c^2} = \frac{40}{511} = 0.078$$

$$\beta^2 = 1 - \frac{1}{(1+\tau)^2} = 1 - \frac{1}{(1+0.078)^2} = 0.139$$

$$\eta = \eta_c \times 1.7 \times 10^{-5} Z^{2/3} \frac{1}{\tau(\tau+2)} = 1.198 \times 1.7 \times 10^{-5} \times 7.42^{2/3} \frac{1}{0.078 \times (0.078+2)}$$

$$= 4.78 \times 10^{-4}$$

$$\sigma_{el} = \pi Z(Z+1) r_e^2 \frac{1-\beta^2}{\beta^4} \frac{1}{\eta(\eta+1)}$$

$$= \pi \times 7.42 \times (7.42+1)(2.8179 \times 10^{-13})^2 \frac{1-0.139}{0.139^2}$$

$$\times \frac{1}{4.78 \times 10^{-4} \times (4.78 \times 10^{-4} +1)} = 1.45 \times 10^{-18} \, \text{cm}^2$$

13.4 Stopping Powers

To examine the reliability of inelastic cross sections, a calculation of the collision stopping power composed of ionization and excitation components was performed. This is an exact evaluation without any parameter, such as the mean excitation energy I. Total collision stopping power is composed of three terms: (1) kinetic energy of secondary electrons, (2) potential energy of the subshell, and (3) excitation.

$$S_{col} = S_{kin} + S_{pot} + S_{exc}$$

$$= \rho \frac{N_A}{A} [\sigma_{ion}(\bar{\varepsilon} + \bar{B}) + \sigma_{exc}\bar{E_a}] \tag{13.21}$$

where ρ is the water density ($= 1$ g cm^{-3}), N_A is the Avogadro number, A is the molecular weight of water, and

$$\bar{\varepsilon} = \sum_j \bar{\varepsilon}_j \sigma^{(j)} \bigg/ \sum_j \sigma^{(j)}, \qquad \bar{B} = \sum_j \bar{B}_j \sigma^{(j)} \bigg/ \sum_j \sigma^{(j)} \tag{13.22}$$

The mean energy of secondary electrons emitted from the jth orbital, $\bar{\varepsilon}_j$, is calculated by

$$\bar{\varepsilon}_j = \int_0^{(T-B_j)/2} \varepsilon\left(\frac{d\sigma^{(j)}}{d\varepsilon}\right) d\varepsilon \bigg/ \int_0^{(T-B_j)/2} \left(\frac{d\sigma^{(j)}}{d\varepsilon}\right) d\varepsilon \tag{13.23}$$

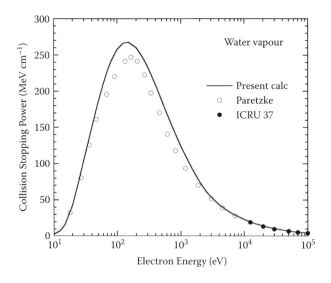

FIGURE 13.10
Comparison of collision stopping powers for water vapor as a function of the impact energy of electrons.

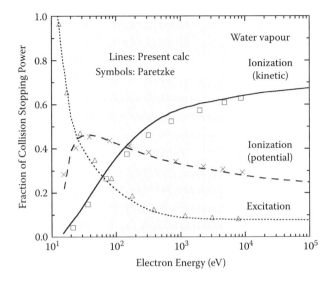

FIGURE 13.11
Fractions of collision stopping powers for water vapor in comparison with those of Paretzke (1988).

Figure 13.10 shows the total collision stopping powers estimated by the present model in comparison with the published data. The stopping powers for less than 10 keV electrons were compared with the calculated data of Paretzke (1988). For above 10 keV, the discrepancies with the values for the collision stopping power tabulated in ICRU 37 (1984) were less than a few percent. Figure 13.11 shows the fractions of each component. These fractions for electron energies 10 eV–10 keV were in agreement with the calculations of Paretzke (1988) and Paretzke and Berger (1978).

13.5 Summary

1. Energy spectra for secondary electrons were calculated using Seltzer's formula. Angular distributions of secondary electrons were calculated using a combination of the experimental data and the kinematical relationship for low- and high-energy electrons.

2. A total ionization cross section was obtained by least squares fitting of various experimental data.

3. Partial excitation cross sections and the transition energies for the major excitation modes compiled by Paretzke were adopted. A total excitation cross section was evaluated by the Fano plot.

4. Elastic scattering cross sections were calculated using the Rutherford formula, taking into account the screening parameter given by Moliere. Angular distributions below 1 keV were taken from various experimental data.

5. Calculations of the collision stopping power composed of ionization and excitation components were performed without any parameters, such as the mean excitation energy and the effective charge.

QUESTIONS

1. What is a Fano plot and what is it used for?
2. What are the sources of uncertainties in Monte Carlo track structure simulations? How could these be estimated?
3. Using the Bragg additivity rule, $S_M = \alpha S_A + \beta S_B + \gamma S_C + ...,$ where S_M is the stopping power of the gas mixture constituents, gases A, B, and C; and α, β, and γ are the fractional compositions by weight of the individual gases, obtain the stopping power of electrons in water, and compare with ICRU values. To what extent can the additivity rule be applied in obtaining stopping power values of mixtures? Discuss the limitation of the method.

References

Berger MJ, Wang R. 1988. Multiple scattering angular deflections and energy-loss straggling. In *Monte Carlo Transport of Electrons and Photons*, ed. TM Jenkins, WR Nelson, A Rindi. New York: Plenum, pp. 21–56.

Bolorizadeh MA, Rudd ME. 1986. Angular and energy dependence of cross sections for ejection of electrons from water vapor. I. 50–2000 eV electron impact. *Phys. Rev. A* 33: 882–887.

Danjo A, Nishimura H. 1985. Elastic scattering of electrons from H_2O molecule. *J. Phys. Soc. Jpn.* 54: 1224–1227.

Djuric NLj, Cadez IM, Kurepa MV. 1988. H_2O and D_2O total ionisation cross-sections by electron impact. *Int. J. Mass Spectrom. Ion Proc.* 83: R7–R10.

Green AES, Stolarski RS. 1972. Analytic models of electron impact excitation cross sections. *J. Atm. Terr. Phys.* 34: 1703–1717.

Hayashi M. 1989. Electron collision cross-sections for atoms and molecules determined from beam and swarm data. *IAEA-TECDOC* 506: 193–199.

Hubbell JH. 1977. Photon mass attenuation and mass energy-absorption co-efficients for H, C, N, O, Ar, and seven mixtures from 0.1 keV to 20 MeV. *Radiat. Res.* 70: 58–81.

ICRU. 1984. *Stopping Powers for Electrons and Positrons*. ICRU Report 37.

Katase A, Ishibashi K, Matsumoto Y, Sakae T, Maezono S, Murakami E, Watanabe K, Maki H. 1986. Elastic scattering of electrons by water molecules over the range 100–1000 eV. *J. Phys. B At. Mol. Phys.* 19: 2715–2734.

Kim YK. 1975. Energy distribution of secondary electrons. I. Consistency of experimental data. *Radiat. Res.* 61: 21–35.

Mann JB. 1968. Atomic structure calculations. II. Hartree-Fock wavefunctions and radial expectation values. *Hydrogen to Lawrencium* LA-3691.

Moliere G. 1948. Theorie der Streuung schneller gelandener Teilchen II: Mehrfach- und Vielfachstreuung. *Z. Naturf.* 3a: 78–97.

Nikjoo H, Uehara S, Emfietzoglou D, Cucinotta FA. 2006. Track-structure codes in radiation research. *Radiat. Measurements* 41: 1052–1074.

Nishimura H. 1979. Elastic scattering cross-sections of H_2O by low energy electrons. In *Electronic and Atomic Collisions, Proceedings of XIth International Conference*, ed. N Oda, K Takayanagi. Amsterdam: North-Holland, p. 314.

Olivero JJ, Stagat RW, Green AES. 1972. Electron deposition in water vapour, with atmospheric applications. *J. Geophys. Res.* 77: 4797–4811.

Opal CB, Beaty EC, Peterson WK. 1972. Tables of secondary-electron-production cross-sections. *At. Data* 4: 209–253.

Paretzke HG. 1988. *Simulation von Elektronenspuren in Energiebereich 0.01–10 keV in Wasserdampf GSF-Bericht 24/88.* Munich: Institut fur Strahlenschutz der Gesellschaft fur Strahlen-und Umweltforschung.

Paretzke HG, Berger MJ. 1978. Stopping power and energy degradation for electrons in water vapour. In *Proceedings of 6th Symposium on Microdosimetry* (Brussels, 1978), ed. J Booz, HG Ebert. London: Harwood, pp. 749–758.

Schutten J, de Heer FJ, Moustafa HR, Boerboom AJH, Kistemaker J. 1966. Gross- and partial-ionisation cross sections for electrons on water vapor in the energy range 0.1–20 keV. *J. Chem. Phys.* 44: 3924–3928.

Seltzer SM. 1988. Cross sections for bremsstrahlung production and electron-impact ionization. In *Monte Carlo Transport of Electrons and Photons*, ed. TM Jenkins, WR Nelson, A Rindi. New York: Plenum, pp. 81–114

Shyn TW, Cho SY. 1987. Vibrationally elastic scattering cross section of water vapor by electron impact. *Phys. Rev.* A36: 5138–5142.

Siegbahn K. 1969. *ESCA Applied to Free Molecules.* Amsterdam: North-Holland.

Trajmar S, Williams W, Kuppermann A. 1973. Electron impact excitation of H_2O. *J. Chem. Phys.* 58: 2521–2531.

Uehara S, Nikjoo H, Goodhead DT. 1999. Comparison and assessment of electron cross sections for Monte Carlo track structure codes. *Radiat. Res.* 152: 202–213.

Vroom DA, Palmer RL. 1977. Measurement of energy distributions of secondary electrons ejected from water vapour by fast electrons. *J. Chem. Phys.* 66: 3720–3723.

Yeh JJ, Lindau I. 1985. Atomic sub-shell photoionization cross-sections and asymmetry parameters: 1<Z<103. *At. Data Nucl. Data Tables* 32: 1–155.

Zaider M, Brenner DJ, Wilson WE. 1983. The applications of track calculations to radiobiology. I. Monte Carlo simulation of proton tracks. *Radiat. Res.* 95: 231–247.

14

Cross Sections for Interactions of Low-Energy Protons (<1 MeVu^{-1}) in Water

14.1 Ionization

14.1.1 Secondary Electrons

For fast ions, the majority of energy is transferred in ionizing collisions, resulting in energetic free electrons and the potential energy of residual ions. Energy spectra of secondary electrons ejected by the proton (H$^+$) impact were calculated using the empirical model given by Rudd (1988):

$$\frac{d\sigma_i(\varepsilon)}{d\varepsilon} = \frac{S}{B} \frac{F_1 + F_2 w}{(1+w)^3 \{1 + \exp[\alpha(w - w_c)/v]\}} \tag{14.1}$$

where $S = 4\pi a_0^2 N z^2 (R/B)^2$, ε = electron energy, B = binding energy for each orbital, $a_0 = 0.529 \times 10^{-8}$ cm, N = electron number in orbital, z = charge of the projectile nucleus, $R = 13.6$ eV, T = kinetic energy of protons in eVu^{-1} units, $\lambda = m_p/m_e = 1{,}836$, $w = \varepsilon/B$, $v = (T/\lambda B)^{1/2}$, and $w_c = 4v^2 - 2v - R/4B$. F_1, F_2, and α are adjustable fitting parameters.

$$F_1(v) = L_1 + H_1$$

$$L_1 = C_1 v^{D_1}/(1 + E_1 v^{D_1+4}) \tag{14.2a}$$

$$H_1 = A_1 \ln(1 + v^2)/(v^2 + B_1/v^2)$$

$$F_2(v) = L_2 H_2/(L_2 + H_2)$$

$$L_2 = C_2 v^{D_2} \tag{14.2b}$$

$$H_2 = A_2/v^2 + B_2/v^4$$

Table 14.1 lists the fitting parameters for water vapor given by Rudd et al. (1992).

TABLE 14.1

Fitting Parameters for Rudd's Formula

	A_1	A_2	B_1	B_2	C_1	C_2	D_1	D_2	E_1	α
Outer shell	0.97	1.04	82.00	17.30	0.40	0.76	−0.30	0.04	0.38	0.64
K-Shell	1.25	1.10	0.50	1.30	1.00	1.00	1.00	0.00	3.00	0.66

Source: Values taken from Rudd, M.E., et al., *Rev. Mod. Phys.*, 64, 441–490, 1992.

   This model gives the single differential cross sections (SDCSs) for second-ary electron production by proton impact, and also the partial ionization cross sections for each orbital. The boundary energy for energetic second-ary electrons generated by ionization was determined as 1 eV. If electron energy is <1 eV, then the energy deposition is $\Delta E = B_i + \varepsilon$ for the ith orbital ionization. The accuracy of the Rudd model was confirmed by comparing it with the experimental spectra taken from Bolorizadeh and Rudd (1986a) for $T = 20$, 50, and 150 keV and from Wilson et al. (1984) for $T = 1$ MeV. It is assumed that the spectra of electrons ejected from the target molecule by neutral hydrogen (H^0) impact are equal to those for protons. Figure 14.1 shows the calculated SDCS of ejected electrons for various energy of H^+. Calculations take into account the relative intensity of the partial ioniza-tion cross sections.

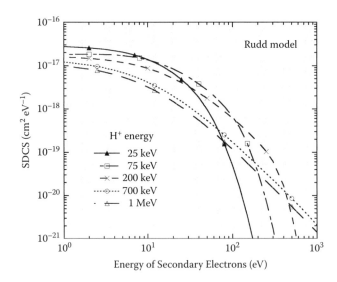

FIGURE 14.1
Calculated energy spectra for secondary electrons ejected by H^+ impact with energies of 10, 50, 100, and 500 keVu^{-1} and 1 MeVu^{-1}.

EXAMPLE 14.1

Write down a formula for calculating the mean electron energy emitted by ionization of a water molecule by proton impact. Take into account different electron energy spectra for different orbitals.

SOLUTION 14.1

$$\bar{\varepsilon} = \sum_{l=1}^{5} P_l \int \varepsilon \frac{d\sigma_l(\varepsilon)}{d\varepsilon} d\varepsilon$$

where P_l is the partial cross section of the orbital l, calculated from

$$P_l = \frac{\int \dfrac{d\sigma_l(\varepsilon)}{d\varepsilon} d\varepsilon}{\sum\limits_{l'=1}^{5} \int \dfrac{d\sigma_{l'}(\varepsilon)}{d\varepsilon} d\varepsilon}$$

Angular distributions of electrons ejected from water by H⁺ impact are measured by Bolorizadeh and Rudd (1986a) and Toburen and Wilson (1977) in the energy range between 15 keV and 1 MeV. For random sampling of the angular distributions of ejected electrons, we directly used the experimental data for various electron energies at $T = 15, 100, 300,$ and 500 keV, and 1 MeV. Interpolation was made for the intermediate energies of ejected electrons and incident protons. Again, we assumed the double differential cross section (DDCS) for the H⁰ impact to be equal to that of H⁺. Figure 14.2 shows the randomly sampled DDCS normalized to the absolute experimental values at $T = 15$ keVu⁻¹ (a) and 300 keVu⁻¹ (b).

ICRU (1996) recommended the semiempirical formula for the DDCS called the Hansen-Kocbach-Stolterfoht (HKS) model, which appears to be promising. Hansen and Kocbach (1989) derived a simple expression for the DDCS as a function of the impact parameter.

$$\frac{d^2\sigma_i(\varepsilon,\theta)}{d\varepsilon d\Omega} = \frac{16a_0^2 Z_1^2}{3\pi R v_1^2 \alpha k_c^3} \left[\frac{\alpha_c^4}{\alpha_c^4 + (K_m - k_t \cos\theta)^2} \right]^3$$

$$k_c = \sqrt{k^2 + \frac{2\alpha^2}{\ln\left(2v_1^2/\alpha^2\right)}},$$

$$k_t = \sqrt{k^2 + 0.2\alpha^2 \sqrt{v_1/\alpha}},$$

$$\alpha_c = \alpha\left(1 + 0.7 \frac{v_1^2}{v_1^2 + k^2}\right),$$

(14.3)

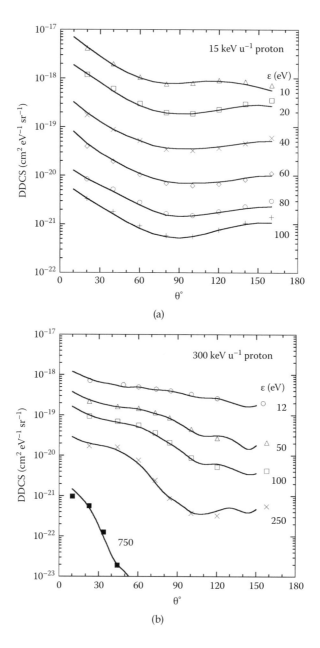

FIGURE 14.2
(a) Angular distributions of secondary electrons with various energies ε (eV) ejected by 15 keVu^{-1} proton impact. Lines: present calculations; symbols: experimental data. (From Bolorizadeh, M.A., and Rudd, M.E., *Phys. Rev.* A33, 888–892, 1986a.) (b) Angular distributions of secondary electrons with various energies ε (eV) ejected by 300 keVu^{-1} proton impact. Lines: present calculations; symbols: experimental data. (From Toburen, L.H., and Wilson, W.E., *J. Chem. Phys.*, 66, 5202–5213, 1977.)

where Z_1 is the charge, $v_1 = (T/\lambda R)^{1/2}$ the velocity of the projectile, and $K_m = (a^2 + k^2)/2v_1$ is the minimum momentum transfer. The quantities k_c and k_t are small modifications of the quantity $k = (\varepsilon/R)^{1/2}$, which is the momentum of the outgoing electron. α_c is a modification of the mean initial momentum parameter $\alpha = (B/R)^{1/2}$. This equation is for a single target electron and must be summed over all the electrons in the target. The experimental DDCS was compared with the theoretical one evaluated by the HKS model. The model was in good agreement with the experimental data for high-energy protons. However, the reproducibility for lower than 100 keV protons was unsatisfactory. Therefore, it is realistic to use the experimental and interpolated data.

EXAMPLE 14.2

Explain the shoulder of the DDCS for ionization by 300 keV proton impact (Figure 14.2) for the electron energy 250 eV.

SOLUTION 14.2

The collision of the proton and the orbital electron can be seen as a hard collision in this case (relatively large energy transferred to the electron). By neglecting the binding energy of the electron, the emission angle can be estimated using the kinematic relationship, Equation (10.1), as follows:

$$\cos\theta = \sqrt{\frac{m_{HI}\varepsilon}{4m_e T}}$$

The proton-to-electron mass ratio is 1,836. Therefore, for 250 eV electrons $\theta = \arccos(\sqrt{\frac{1836 \cdot 250}{4 \cdot 300 \cdot 10^3}}) = 51.80°$, close to the angle where the DDCS curve exhibits a shoulder (Figure14.2b).

14.1.2 Total Cross Sections

Total ionization cross sections (TICSs) for protons $\sigma_{ip}(T)$ were determined by the least squares fit to the experimental data given by Rudd et al. (1985), in which the six data sets measured under the different experimental conditions are provided. These six data sets fall into two groups: two sets are larger in magnitude than the other four. Only the four sets belonging to the smaller group consisting of low-energy accelerator, early Van de Graaff data, early UNL data, and late UNL data gave consistent agreement with published stopping cross sections in the high-energy regions, and were therefore used for subsequent calculations. The four sets were fitted by power functions at the lower- and higher-energy regions (Uehara et al. 2001).

$$\sigma_{ip}(T) = 10^{\alpha} \tag{14.4}$$

$$\alpha = -0.24653 + 0.68546 \log T + 0.11287(\log T)^2 - 0.11194(\log T)^3$$

$$\text{for} \quad 1 \le T \le 100 \text{ keV}$$

$$\alpha = -3.96257 + 6.08376 \log T - 2.46783(\log T)^2 + 0.29425(\log T)^3$$

$$\text{for} \quad 100 \text{keV} < T \le 1 \text{MeV}$$

where log is to base 10. The cross sections from the fits have units of 10^{-16} cm^2 with T in keV.

The total ionization cross section for the neutral hydrogen impact was calculated by an analytical function developed by Green and McNeal (1971):

$$\sigma_{iH}(T) = \frac{\sigma_0 (Za)^{\Omega}(T-B)^{\nu}}{J^{\Omega+\nu} + T^{\Omega+\nu}} \tag{14.5}$$

where $\sigma_0 = 10^{-16}$ cm^2, and the parameters are given in Table 3 of the literature: $B = 0.0126$, $Z = 2$, $a = 47.45$, $J = 23.68$, $\Omega = 0.75$, and $\nu = 1.04$ for H$_2$, and $B = 0.0126$, $Z = 16$, $a = 73.88$, $J = 73.11$, $\Omega = 0.75$, and $\nu = 0.43$ for O$_2$. Cross sections for H$_2$O were obtained from the additivity relationship $\sigma(H_2) + \sigma(O_2)/2$ using the calculated values for H$_2$ and O$_2$. Figure 14.3 shows the TICSs for H$^+$ and H^0 in the energy range between 1 keV^{-1} and 1 MeVu^{-1}.

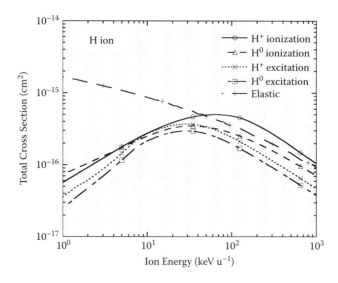

FIGURE 14.3
Total cross sections due to H$^+$ and H^0 impact on water vapor.

14.2 Excitation

There are no experiments on excitation cross sections for proton impact for water. The proton cross sections obtained by scaling of the electron excitation cross sections are plausible for high-energy protons of >500 keV (Dingfelder et al. 2000). To cover the lower-energy regions, the semiempirical model developed by Miller and Green (1973) was adopted, which is based on the electron impact excitation. The analytical form for an excitation cross section $\sigma_e(T)$ is the same as Equation (14.5), but replacing the ionization threshold B by energy of the excited state, W, and the values of parameters. The parameters for 28 excitation states are given in Table III of the literature. In the work of Miller and Green, cross sections for excitation by neutral hydrogen were calculated using the same values as for protons except the parameter a of the formula was assumed to be three-quarters of the proton value. The overall magnitude becomes ~80% of the proton cross sections, though the shape is analogous. Miller and Green noted that their excitation cross sections were only rough estimates proposed in the absence of detailed theoretical or experimental data. They further noted that the position and height of the maximum in the cross section could easily be in error by a factor of two and that the nature of the low-energy fall-off was speculative. The mean excitation energy W_{av} was calculated using the relationship $\Sigma W_j \sigma_{ej}(T)/\Sigma \sigma_{ej}(T)$, where W_j is the excitation threshold energy and $\sigma_{ej}(T)$ is the excitation cross section for state j. The values of W_{av} increase slowly from 12 eV at 1 keV to around 14 eV for 1 MeV. Figure 14.3 shows the total excitation cross sections for H^+ and H^0 in the energy range between 1 keV⁻¹ and 1 MeVu⁻¹.

14.3 Elastic Scattering

High-energy ions proceed along a straight-line trajectory. However, the effects of elastic nuclear collisions will become more prominent in the low-energy region. Elastic collisions transfer little or no energy but can have a significant effect on the spatial character of the track structure at very low energies. Also, the nuclear energy loss cannot be neglected at <10 keVu⁻¹. Nuclear elastic scattering was evaluated using the classical mechanics trajectory calculations (CMTCs) taking into account the screening by atomic electrons (ICRU 1993). The interaction between the projectile and the target atom can be described in terms of a potential function:

$$V(r) = (zZe^2/r)F_s(r/r_s) \tag{14.6}$$

The first factor in this equation is the Coulomb potential for two bare nuclei with charges ze and Ze. The factor $F_s(r/r_s)$ takes into account the screening by atomic electrons. The extent of screening is characterized by the screening length, r_s. A commonly used prescription for the screening length is

$$r_s = 0.88534r_B(z^{2/3} + Z^{2/3})^{-1/2} \qquad (14.7)$$

where r_B is the Bohr radius and 0.88534 is a numerical constant from the Thomas-Fermi model of the atom. For protons, a screening function $F_s(r/r_s)$ is given by (ICRU 1993)

$$F_s(r/r_s) = 0.10e^{-6r/r_s} + 0.55e^{-1.2r/r_s} + 0.35e^{-0.3r/r_s} \qquad (14.8)$$

EXAMPLE 14.3

Compare the screening lengths for the collisions of H^+ and C^{6+} with water.

SOLUTION 14.3

Assume $Z = 10$ for water, the screening lengths of the target nucleus:

For H^+ impact: $r_s = \dfrac{0.88534 \cdot 5.29 \cdot 10^{-9} \text{ cm}}{\sqrt{1^{2/3} + 10^{2/3}}} = 1.971 \cdot 10^{-9} \text{ cm}$

For C^{6+} impact: $r_s = \dfrac{0.88534 \cdot 5.29 \cdot 10^{-9} \text{ cm}}{\sqrt{6^{2/3} + 10^{2/3}}} = 1.662 \cdot 10^{-9} \text{ cm}$

The screening is less effective for the carbon ion.

For a particle scattered in a central potential $V(r)$, the deflection angle θ is obtained in a classical mechanics trajectory calculation as a function of the impact parameter, p (Mott and Massey 1965):

$$\theta = \pi - 2\int_{r_{min}}^{\infty} \frac{p}{r^2\sqrt{1 - V(r)/T_{cm} - p^2/r^2}} \, dr \qquad (14.9)$$

where r_{min} is the distance of closest approach. This value is given by the positive root of the following:

$$r^2 - 2zZr_e \frac{m_e c^2}{\beta Pc} rF_s(r/r_s) - p^2 = 0 \qquad (14.10)$$

in which

$$\beta = \sqrt{1 - \left(\frac{m_p c^2}{T + m_p c^2}\right)^2}, \qquad Pc = \sqrt{(T + m_p c^2)^2 - (m_p c^2)^2} \qquad (14.11)$$

and $r_e = 2.81794 \times 10^{-13}$ cm, $m_e c^2$ = electron rest mass, $m_p c^2$ = proton rest mass, and P is the momentum of proton. T_{cm} is the particle energy in the center of mass system, given by

$$T_{cm} = \frac{T}{1 + m_p/M_t} \qquad (14.12)$$

where T is the energy in the laboratory system, and m_p and M_t are the masses of proton and target atom. Equation (14.9) was solved numerically using procedures given by Everhart et al. (1955) and Wilson and Haggmark (1977). Thereby the deflection angle θ was obtained as a function of the impact parameter p. The boundary of large-angle calculations and small-angle calculations described by Everhart et al. was set at $\theta = 0.1\pi$, where smooth transitions are given.

The differential elastic scattering cross section can be obtained by numerical differentiation of the curve of impact parameter vs. deflection angle using a B-spline interpolation (Goldstein 1950):

$$\frac{d\sigma_{el}}{d\Omega} = -\frac{p}{\sin\theta}\frac{dp}{d\theta} \qquad (14.13)$$

θ and $d\sigma_{el}/d\Omega$ are functions of the impact parameter and the particle energy.

In order to reduce the divergence of the total cross section, the cutoff angles, θ_{cut}, were set such as to limit the increase in the scattering probability at low scattering angles (ICRU 1978).

$$\theta_{cut} \approx \frac{1}{137} Z^{1/3} \frac{m_e c^2}{Pc} \qquad (14.14)$$

The total elastic scattering cross section for the proton energy of T is calculated by

$$\sigma_{el}(T) = 2\pi \int_{\theta_{cut}}^{\pi} \frac{d\sigma_{el}}{d\Omega} \sin\theta d\theta \qquad (14.15)$$

The minimum impact parameter p_{min} was fixed at 0.01, which corresponds to the backward scattering angle ~π. The maximum value p_{max}, which corresponds to the forward scattering, was chosen to reproduce the nuclear stopping powers given in the ICRU database. At the same time, the total cross sections were obtained for this p_{max} value. Figure 14.3 shows the total cross section (Equation (14.15)) of proton elastic scattering in water vapor. Figure 14.4 shows the p_{max} and the values of θ_{cut} (Equation (14.14)) as a function of the proton energy. Figure 14.5 shows examples of the angular distributions of elastic scattering at small angles for the proton energies of 1, 10, and 100 keV and 1 MeV. Scattering angle θ is randomly sampled by a

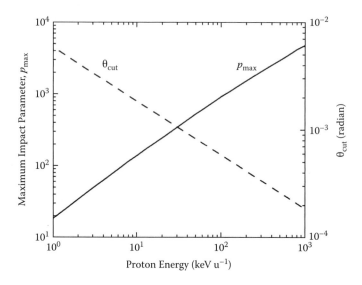

FIGURE 14.4
Maximum impact parameter and the cutoff angle for proton elastic scattering in water.

table storage and lookup method in the Monte Carlo track code. The table was predigested by separated data preparation routines based on the direct sampling method using differential cross sections (Equation (14.13)). We assumed the elastic scattering cross sections for neutral hydrogen (H^0) to be equal to those for protons.

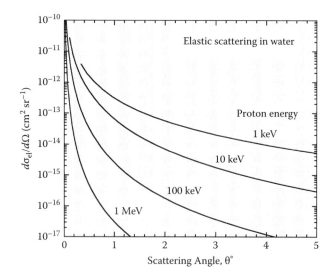

FIGURE 14.5
Differential cross sections of elastic scattering in water for H^+ and H^0 as a function of scattering angle θ.

14.4 Charge Transfer

When fast ions (ions with energy in the MeV region) slow down around the Bragg peak (~0.3 $MeVu^{-1}$), interactions involving electron capture and loss by the moving ions become an increasingly important component of the energy loss process. Table 14.2 presents a list of charge transfer interactions for low-energy H^+ and H^0 projectiles with water. For incident protons with speeds comparable with or lower than the orbital speeds of the bound electrons in the target, the capture of electrons from the target is a significant mechanism that ionizes the target but does not produce secondary electrons.

If the projectile is a neutral atom or a dressed ion (i.e., an ion bearing one or more electrons), a collision can eject electrons from either the projectile or the target. For a fast projectile, one can consider the projectile a simple collection of a nucleus and electrons traveling together at the same speed. After the charge transfer process the neutral hydrogen atom either becomes stripped or ionizes a water molecule. In the first case, a proton and a possibly excited water molecule remain, and an electron is ejected in the forward direction with nearly the same velocity as the proton. In the latter case, the neutral hydrogen, an ionized water molecule, and an ejected electron with a similar distribution of electron energy and ejection angle as that for protons result. This case is equal to the ionization process by H^0 impact. Charge transfer can also be accompanied by simultaneous target ionization for both electron loss and electron capture processes. However, there is insufficient information on the cross sections for these interactions to be able to be included in these calculations. Toburen (1998) gives a fine review on the basic data of interactions for low-energy ions with various mediums focusing on ionization and charge transfer.

Charge transfer cross sections are generally designated as σ_{if}, where i is the initial and f the final charge state of the moving particle. To develop track structure models, cross sections are needed for both electron capture and electron loss. Charge transfer cross sections for protons were compiled based on available experimental data for H^+ and H^0. Total cross sections for electron capture σ_{10} and electron loss σ_{01} for protons and neutral hydrogen, respectively, were evaluated using the analytical functions developed by

TABLE 14.2

Charge transfer for low energy H^+ and H^0 projectiles with water.

electron capture	$H^+ + H_2O \rightarrow H^0 + H_2O^+$	(σ_{10})
electron loss	$H^0 + H_2O \rightarrow H^+ + H_2O + e$	(σ_{01})
electron capture and target ionization	$H^+ + H_2O \rightarrow H^0 + H_2O^{2+} + e$	
electron loss and target ionization	$H^0 + H_2O \rightarrow H^+ + H_2O^+ + 2e$	

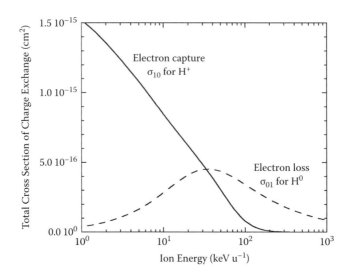

FIGURE 14.6
Charge exchange cross sections due to H ion impact on water.

Miller and Green (1973) and fitted to the experimental data of Toburen et al. (1968) and Dagnac et al. (1970). Electron capture is fitted by the equation as

$$\sigma_{10} = \sigma_0 \left[\left(\frac{T}{A} \right)^p + \left(\frac{T}{B} \right)^q + \left(\frac{T}{C} \right)^r \right]^{-1} \tag{14.16}$$

where $\sigma_0 = 2 \times 10^{-15}$ cm^2, $A = 44.1$ keV, $B = 6.0$ keV, $C = 1.5$ keV, $p = 3.52$, $q = 0.667$, and $r = -0.027$. Electron loss σ_{01} is fitted by the same formula as Equation (14.5), but with the following parameters: $\sigma_0 = 10^{-16}$ cm^2, $Z = 10$, $a = 79.3$ keV, $B = 12.6$ eV, $J = 27.7$ keV, $\Omega = 0.652$, and $v = 0.943$. Figure 14.6 shows the charge transfer cross sections for the proton impact.

14.5 Stopping Powers

14.5.1 Electronic Stopping Powers

In order to test the consistency of interaction cross sections, we have evaluated the electronic stopping powers from the basic interaction cross sections for the different interactions noted above without introducing an effective charge of the moving ions (Uehara et al. 2000). The electronic stopping power for protons is obtained by

$$S_e = \rho \frac{N_A}{A} \sigma_{stop} \tag{14.17}$$

in which stopping cross section, σ_{stop}, including the contribution from neutral hydrogen, is defined by

$$\sigma_{\text{stop}} = F_+\left(Q_{\text{ion}}\sigma_{i+} + Q_{\text{exc}}\sigma_{e+} + Q_{10}\sigma_{10}\right) + F_0\left(Q_{\text{ion}}\sigma_{i0} + Q_{\text{exc}}\sigma_{e0} + Q_{01}\sigma_{01}\right) \qquad (14.18)$$

where σ represents total cross sections; the subscripts $+$ and 0 represent H^+ and H^0, respectively; Q is the mean energy transfer in each interaction; and F_+ and F_0 are the equilibrium charge fractions for protons and neutral hydrogen, respectively. All data are a function of projectile kinetic energy T.

Since low-energy protons and neutral hydrogen are continuously undergoing the charge exchange process, equilibrium will exist between the two charge states. For a two-component system consisting of only positive and neutral charge states, the fractions of hydrogen and protons in the equilibrium charge state are given by the conventional formula

$$F_0 = \sigma_{10}/(\sigma_{10} + \sigma_{01}) \quad \text{and} \quad F_+ = 1 - F_0 \qquad (14.19)$$

The mean energy transfer for ionization by proton impact is

$$Q_{\text{ion}} = \varepsilon_{\text{av}} + B_{\text{av}} \qquad (14.20)$$

where ε_{av} is the average energy of secondary electrons produced and B_{av} is the weighted mean of the binding energy of the ejected electron such as $\Sigma f_i B_i$. These values can be obtained by the Rudd model (Equation (14.1)). For ε_{av} values, the experimental data are available at several proton energies. Bolorizadeh and Rudd (1986a) have given for the proton energies between 15 and 150 keV, and Toburen et al. (1980) have presented at 300 and 500 keV from equivalent velocity He^+ data. Finally, taking into account these experimental data, ε_{av} was determined. The mean energy transfer for excitation Q_{exc} is equal to W_{av}, which has been described in Section 14.2.

EXAMPLE 14.4

Show that Equation (14.19) is valid under the equilibrium condition of the charge exchange processes.

SOLUTION 14.4

At equilibrium the rate of electron capture and loss of a charge state is the same. Therefore,

$$\sum_j (F_j\sigma_{ji} - F_i\sigma_{ij}) = 0 \quad \text{and} \quad \sum_j F_j = 1$$

where F_j is the fraction of the charge state j, and σ_{ji} is the cross section for charge exchange where the charge state j is captured or loses electrons and becomes the charge state i. For hydrogen, there are two charge states, $j = 0, 1$, and it is reasonable to assume only a single electronic charge exchange, σ_{01} and σ_{10}; thus:

$$(F_0\sigma_{01} - F_1\sigma_{10}) = 0 \Rightarrow F_0 = \frac{F_1\sigma_{10}}{\sigma_{01}} = \frac{(1-F_0)\sigma_{10}}{\sigma_{01}} \Rightarrow F_0 = \frac{\sigma_{10}}{\sigma_{01} + \sigma_{01}}$$

and

$$F_1 = 1 - F_0 = \frac{\sigma_{01}}{\sigma_{01} + \sigma_{01}}$$

Three energy loss components were taken into account for energy transfer in the electron capture process. These include (1) energy to remove the electron from the target molecule, (2) energy to provide translational velocity for the electron to move at the speed of the proton, and (3) energy released from capture into the bound state of the proton. The latter is equal to the binding potential of the state to which the electron is captured. Therefore, the net energy lost by the proton in capturing an electron is given by

$$Q_{10} = \frac{1}{2}m_e v_p^2 + B_{H_2O} - B_H = \frac{T}{\lambda} + B_{H_2O} - B_H \tag{14.21}$$

where m_e is the electron mass, m_p is the proton mass, v_p is the proton velocity, λ is $m_p/m_e = 1{,}836$, and B_{H2O} (B_{av}) and B_H (13.6 eV) are the binding energies of the target and projectile, respectively. The energy lost by the neutral hydrogen atom in the electron loss process is simply the binding energy of the projectile electron. The kinetic energy of the electron resulting from the electron loss process was provided for the capture process. Thus, one can simply write the energy lost by the neutral projectile in this process as

$$Q_{01} = B_H \tag{14.22}$$

Figure 14.7 presents the mean energy transfer obtained for ionization, excitation, and charge exchange. We used the same values of mean energy transfer for ionization and excitation for both proton and hydrogen impact. Figure 14.8 shows the calculated fractions of electronic stopping power as a function of proton energy in which contributions from both proton and neutral hydrogen interactions are included. Above 50 keV the kinetic energy transferred to ejected secondary electrons is the largest part of the stopping power. The fraction of 0.70 agrees well with the experimental fraction 0.67 ± 0.04 for 1 MeV protons, which has been estimated by Toburen and Wilson (1977) and Lynch et al. (1976). Charge exchanges become significant

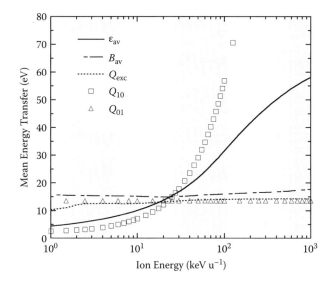

FIGURE 14.7
Mean energy transfers in collisions of H$^+$ and H^0 with water vapor leading to emission of secondary electrons or excitations and charge transfer processes.

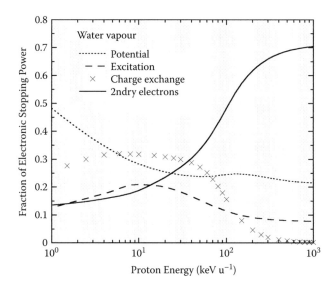

FIGURE 14.8
Fractions of electronic stopping power of water vapor for protons.

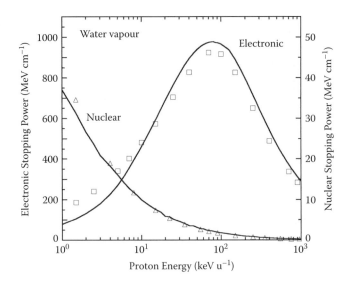

FIGURE 14.9
Comparison of the electronic and nuclear stopping powers of water vapor for protons between the present calculations and the ICRU data. Lines: present calculations; symbols: ICRU Report 49 (1993).

at <30 keV. Figure 14.9 shows the total electronic stopping powers of water vapor in comparison with the published data. The electronic stopping powers are consistent with published data to within ~10% for energies of <1 MeV. It is difficult to identify reasons for discrepancies because of lack of experimental cross sections for many of the interactions, especially at lower energies of <10 keV.

From the track structure viewpoint, it is noted that the stripped electron is regarded as the real secondary electron. In fact, the stripped electron is observed in the spectrum of secondary electrons produced in neutral hydrogen–H_2O interactions as a peak centered at $\varepsilon = T/\lambda$ (Wilson and Toburen 1973; Bolorizadeh and Rudd 1986b). From such considerations, we adopted another energy deposition model in charge exchange processes in which, while keeping the energy balance, the energy deposited ΔE in electron capture (Q_{10}) was assumed to be equal to the average binding energy of a water molecule

$$\Delta E = B_{av} \tag{14.23}$$

In the electron loss process (Q_{01}), an electron is generated with the energy

$$\varepsilon = T/\lambda \tag{14.24}$$

and emitted with the same direction as the neutral hydrogen.

EXAMPLE 14.5

Sketch the scheme explaining the energy deposited and transferred in the charge exchange processes of a proton and neutral hydrogen.

SOLUTION 14.5

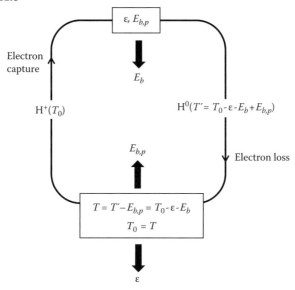

- Thick arrow pointing inwards to the cycle: Energy deposited to the medium.
- Thick arrow pointing outwards to the cycle: Energy transferred to other particles (in this case, an electron).
- Box: Energy that is available (not deposited) in the processes.
- $\varepsilon = T/\lambda$, $E_b = B_{av}$, and $E_{b,p} = 13.6$ eV (the binding energy of the electron to a hydrogen nucleus).

EXAMPLE 14.6

Make an approximate estimation of uncertainties in the electronic stopping cross section.

SOLUTION 14.6

The electronic stopping cross section can be written as

$$L = \sum_{j=1}^{3} f_p \sigma_j Q_j + \sum_{k=1}^{3} f_H \sigma_k Q_k$$

$$(\Delta L)^2 = \left(\sum_{j=1}^{3} f_p \sigma_j Q_j\right)^2 \left(\frac{\Delta f_p}{f_p}\right)^2 + \sum_{j=1}^{3} (f_p \sigma_j Q_j)^2$$

$$\times \left[\left(\frac{\Delta \sigma_j}{\sigma_j}\right)^2 + \left(\frac{\Delta Q_j}{Q_j}\right)^2\right] + \left(\sum_{k=1}^{3} f_H \sigma_k Q_k\right)^2$$

$$\times \left(\frac{\Delta f_H}{f_H}\right)^2 + \sum_{k=1}^{3} (f_H \sigma_k Q_k)^2 \left[\left(\frac{\Delta \sigma_k}{\sigma_k}\right)^2 + \left(\frac{\Delta Q_k}{Q_k}\right)^2\right]$$

where

$$\left(\frac{\Delta f_p}{f_p}\right) = f_H \sqrt{\left(\frac{\Delta \sigma_{10}}{\sigma_{10}}\right)^2 + \left(\frac{\Delta \sigma_{01}}{\sigma_{01}}\right)^2}$$

$$\left(\frac{\Delta f_H}{f_H}\right) = f_p \sqrt{\left(\frac{\Delta \sigma_{10}}{\sigma_{10}}\right)^2 + \left(\frac{\Delta \sigma_{01}}{\sigma_{01}}\right)^2}$$

The relative uncertainties $\Delta\sigma/\sigma$ and $\Delta Q/Q$ are for the total cross section and mean energy transfers for each interaction. Therefore, the relative uncertainty $\Delta L/L$ can be evaluated by inserting the individual uncertainties for $\Delta\sigma/\sigma$ and $\Delta Q/Q$. However, it is difficult to estimate these values rigorously. Estimations of the relative uncertainties in cross section, mean energy transfers, and $\Delta L/L$ are given in the following table.

Solution 14.6

Interactions	$\frac{\Delta\sigma}{\sigma}$ (%)	$\frac{\Delta Q}{Q}$ (%)
p-Ionization	10	10
p-Excitation	50	5
e-Capture	30	5
H-Ionization	20	10
H-Excitation	50	5
e-Loss	30	5
Proton energy (keV)	$\frac{\Delta L}{L}$ (%)	
1	16.5	
10	17.7	
100	14.6	
1,000	13.3	

14.5.2 Nuclear Stopping Powers

The energy transfer to the recoiling target in elastic scattering, $W(\theta,T)$, depends on both the scattering angle and the proton energy. $W(\theta,T)$ is represented by

$$W(\theta,T) = 4T \frac{m_p M_t}{(m_p + M_t)^2} \sin^2 \frac{\theta}{2} \tag{14.25}$$

where m_p and M_t are the masses of the proton and target molecule. The nuclear stopping power S_n at the proton energy T is calculated using the elastic scattering cross section and $W(\theta,T)$.

$$S_n = \rho \frac{N_A}{A} 2\pi \int_{\theta_{cut}}^{\pi} \left(\frac{d\sigma_{el}}{d\Omega} \right) W(\theta,T) \sin\theta \, d\theta \tag{14.26}$$

Figure 14.9 shows the calculated nuclear stopping powers of water vapor in comparison with the published data. The nuclear stopping powers agreed well with the ICRU data (1993).

14.6 Summary

1. Energy spectra of secondary electrons ejected by H^+ and H^0 impact were calculated using the empirical model of Rudd.

2. Experimental angular distributions of electrons were compiled. Interpolation was made where data are lacking.

3. Total ionization cross sections for H^+ were obtained by fitting polynomial functions to the experimental data.

4. Total ionization cross sections for H^0 were calculated by an analytical function of Green and McNeal.

5. Excitation cross sections and mean excitation energy loss were treated based on the formula given by Miller and Green.

6. Nuclear elastic scattering was evaluated using the classical mechanics trajectory calculations.

7. Charge transfer cross sections were compiled based on available experimental data.

8. Calculations of the electronic and the nuclear stopping power were performed without any parameters, such as the mean excitation energy and the effective charge.

QUESTIONS

1. What is approximately the maximum proportion a proton can transfer of its kinetic energy to an electron?
2. Write down an approximate expression for the SDCS for ejection of electrons with kinetic energy between w and $w + d$ by a bare ion with velocity v and charge z. Explain each term of the expression and limits of applicability.
3. Make a brief description of binary encounter formalism (BEF).
4. Total inelastic, ionization, and excitation cross sections are related by $\sigma_T = \sigma_{ioniz} + \sigma_{excit}$. Write down expressions for the mean free path λ for any inelastic interaction, and the fraction of interactions that are ionizing.

References

Bolorizadeh MA, Rudd ME. 1986a. Angular and energy dependence of cross sections for ejection of electrons from water vapor. II. 15-150-keV proton impact. *Phys. Rev.* A33: 888–892.

Bolorizadeh MA, Rudd ME. 1986b. Angular and energy dependence of cross sections for ejection of electrons from water vapor. III. 20-150-keV neutral-hydrogen impact. *Phys. Rev.* A33: 893–896.

Dagnac R, Blanc D, Molina D. 1970. A study on the collision of hydrogen ions H_1^+, H_2^+ and H_3^+ with a water-vapor target. *J. Phys.* B3: 1239–1251.

Dingfelder M, Inokuti M, Paretzke HG. 2000. Inelastic-collision cross sections of liquid water for interactions of energetic protons. *Radiat. Phys. Chem.* 59: 255–275.

Everhart E, Stone G, Carbone RJ. 1955. Classical calculation of differential cross section for scattering from a Coulomb potential with exponential screening. *Phys. Rev.* 99: 1287–1290.

Goldstein H. 1950. *Classical Mechanics*. Reading, MA: Addison-Wesley.

Green AES, McNeal RJ. 1971. Analytic cross sections for inelastic collisions of protons and hydrogen atoms with atomic and molecular gases. *J. Geophys. Res.* 76: 133–144.

Hansen JP, Kocbach L. 1989. Ejection angles of fast delta electrons from K-shell ionisation induced by energetic ions. *J. Phys.* B22: L71–L77.

ICRU. 1978. *Basic Aspects of High Energy Particle Interactions and Radiation Dosimetry*. ICRU Report 28.

ICRU. 1993. *Stopping Powers and Ranges for Protons and Alpha Particles*. ICRU Report 49.

ICRU. 1996. *Secondary Electron Spectra from Charged Particle Interactions*. ICRU Report 55.

Lynch DJ, Toburen LH, Wilson WE. 1976. Electron emission from methane, ammonia, monomethylamine, and dimethylamine by 0.25 to 2.0 MeV protons. *J. Chem. Phys.* 64: 2616–2622.

Miller JH, Green AES. 1973. Proton energy degradation in water vapor. *Radiat. Res.* 54: 343–363.

Mott NF, Massey HSW. 1965. *The Theory of Atomic Collisions*, 3rd ed. London: Oxford University Press.

Rudd ME. 1988. Differential cross sections for secondary electron production by proton impact. *Phys. Rev.* A38: 6129–6137.

Rudd ME, Goffe TV, DuBois RD, Toburen LH. 1985. Cross sections for ionization of water vapor by 7–4000-keV protons. *Phys. Rev.* A31: 492–494.

Rudd ME, Kim Y-K, Madison DH, Gay TJ. 1992. Electron production in proton collisions with atoms and molecules: Energy distributions. *Rev. Mod. Phys.* 64: 441–490.

Toburen LH. 1998. Ionization and charge-transfer: Basic data for track structure calculations. *Radiat. Environ. Biophys.* 37: 221–233.

Toburen LH, Nakai MY, Langley RA. 1968. Measurement of high-energy charge-transfer cross sections for incident protons and atomic hydrogen in various gases. *Phys. Rev.* 171: 114–122.

Toburen LH, Wilson WE. 1977. Energy and angular distributions of electrons ejected from water vapor by 0.3–1.5 MeV protons. *J. Chem. Phys.* 66: 5202–5213.

Toburen LH, Wilson WE, Popowich RJ. 1980. Secondary electron emission from ionization of water vapor by 0.3- to 20-MeV He$^+$ and He^{2+} ions. *Radiat. Res.* 82: 27–44.

Uehara S, Toburen LH, Nikjoo H. 2001. Development of a Monte Carlo track structure code for low-energy protons in water. *Int. J. Radiat. Biol.* 77: 139–154.

Uehara S, Toburen LH, Wilson WE, Goodhead DT, Nikjoo H. 2000. Calculations of electronic stopping cross sections for low-energy protons in water. *Radiat. Phys. Chem.* 59: 1–11.

Wilson WD, Haggmark LG. 1977. Calculations of nuclear stopping, ranges, and straggling in the low-energy region. *Phys. Rev.* B15: 2458–2468.

Wilson WE, Toburen LH. 1973. Electron emission in H$_2^+$-H$_2$ collisions from 0.6 to 1.5 MeV. *Phys. Rev.* A7: 1535–1543.

Wilson WE, Toburen LH, Miller JH, DuBois RD. 1984. *Cross Sections Used in Proton-Track Simulations.* ANL-84-28, pp. 54–62.

15

Cross Sections for Interactions of Low Energy α-Particles (<2 MeVu⁻¹) in Water

15.1 Ionization

15.1.1 Secondary Electrons

The average energies ejected by He^{2+} impact at several hundred keVu⁻¹ of ion energies are almost the same as those by proton impact (Bolorizadeh and Rudd 1986; Toburen and Wilson 1977). The similarity of average energies for ejected electrons suggested that the spectral difference between proton impact and α-particle impact is insignificant. Toburen et al. (1980) showed the ratio of single differential cross sections (SDCSs) for electron emission for equal velocity protons and He^{2+} ions are distributed in the range 1.0 ± 0.2 for various projectile energies. As a first approximation, Rudd's model of proton impact was applied to the SDCS for He^{2+} impact ionization (see Section 14.1). The SDCS for He^{++} was obtained by z^2 scaling of proton data. For ion energies lower than 300 keVu⁻¹, the model of Rudd reproduces well the experimental average energies of secondary electrons. As energy of He^{2+} increases, at energies greater than 300 keVu⁻¹, the model underestimates the average energies up to ~30% at 2 MeVu⁻¹. To correct this discrepancy, calculated electron spectra were modified to produce the average energies using suitable scaling factors.

Toburen et al. (1980) have provided the ratios of SDCSs for ionization of water vapor by He^+ to those for He^{2+} as a function of the ejected electron energy. For ejection of low-energy electrons, i.e., small energy loss by the ion, the ratio decreases due to the screening of the nuclear charge by the He^+-bound electron. This ratio is distributed linearly from 0.4 to 1 for $0 < \varepsilon < 90$ eV for various projectile energies between 200 and 500 keVu⁻¹. For ejection of fast electrons, the cross-sectional ratio approaches 1. Thus, the energy spectra for secondary electrons due to He^+ impact were obtained by modifying the Rudd's model, introducing a simple function $g(\varepsilon)$ such that (Uehara and Nikjoo 2002)

$$\frac{d\sigma_i^{He^+}(\varepsilon)}{d\varepsilon} = g(\varepsilon) \frac{d\sigma_i^{He^{2+}}(\varepsilon)}{d\varepsilon} \tag{15.1}$$

in which

$$g(\varepsilon) = 0.4 + \varepsilon/150 \qquad \text{for } \varepsilon < 90 \text{ eV}$$

$$= 1 \qquad \text{for } \varepsilon \geq 90 \text{ eV} \qquad (15.2)$$

Figure 15.1 shows the calculated SDCS of ejected electrons for various energy of He^{2+} (a) and He^{+} (b).

We assumed the spectra of electrons ejected from the target molecule by He^{0} impact to be equal to those for He^{+}.

The double differential cross sections (DDCSs) for electron emission by heavy ion impact can be calculated using the HKS model described in Section 14.1. The calculated DDCSs show a fairly good agreement with the experimental angular distributions for the ejected electrons with various energies for high-energy projectiles. However, the reproducibility for low-energy projectiles is unsatisfactory. Because the theoretical and model fitting do not provide a satisfactory solution, it is then realistic to use random sampling of angular distributions of ejected electrons from the experimental and interpolated data. Available experimental angular distributions of electrons ejected from water by helium ion impact are limited. Toburen et al. (1980) provide the data in the range 75–500 keVu^{-1} He^{+} and He^{2+}. To cover the broad ion energy range, available proton data were used, assuming that differences of angular distributions between different ions are not very large. The proton data by Bolorizadeh and Rudd (1986) (15 and 100 keVu^{-1}) and by Toburen and Wilson (1977) (1 and 1.5 MeVu^{-1}) were used. The angular distribution of ejected electrons is sampled by a table storage and lookup method. Figure 15.2 presents the randomly sampled DDCSs for various electron energies at $T = 300$ keVu^{-1} of He^{2+} (a) and He^{+} (b). Interpolation was done for the intermediate energies of electrons. Again, we assumed the angular distributions of electrons ejected from the target molecule by He^{0} impact to be equal to those for He^{+}.

15.1.2 Total Cross Sections

To obtain total ionization cross sections (TICSs) for He^{2+}, we fitted a polynomial function to the experimental data of Rudd et al. (1985b) (measured between 10 and 300 keVu^{-1}) and data of Toburen et al. (1980) (measured between 75 and 500 keVu^{-1}). At energies below 10 keVu^{-1} smooth extrapolation was made. For energies greater than 500 keVu^{-1}, extrapolation was made by taking into account both the reproducibility of stopping powers and the scaling of proton ionization cross sections. The proton data (Rudd et al. 1985a) are multiplied by 4 to scale according to z^{2}, as suggested by the Rutherford theory (ICRU 1996). It was found that the scaling law holds asymptotically at energies higher than 1 MeVu^{-1}. The rapid decrease in the lower energies reflects the competition between electron capture and direct target ionization. In this competition electron capture is largely the dominant partner.

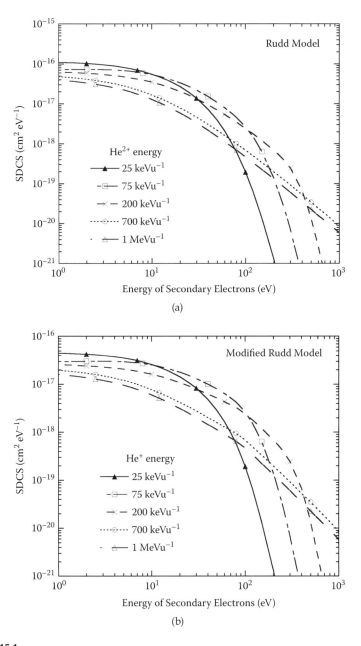

FIGURE 15.1
(a) Calculated energy spectra for secondary electrons ejected by He^{2+} impact with the energies 25, 75, 200, and 700 keVu^{-1} and 1 MeVu^{-1}. (b) Calculated energy spectra for secondary electrons ejected by He^+ impact with the energies 25, 75, 200, and 700 keVu^{-1} and 1 MeVu^{-1}.

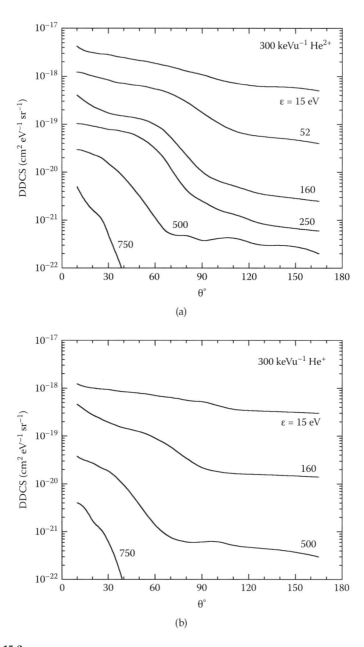

FIGURE 15.2
(a) Angular distributions of secondary electrons for various electron energies ejected by 300 keVu⁻¹ He²⁺ impact. (b) Angular distributions of secondary electrons for various electron energies ejected by 300 keVu⁻¹ He⁺ impact.

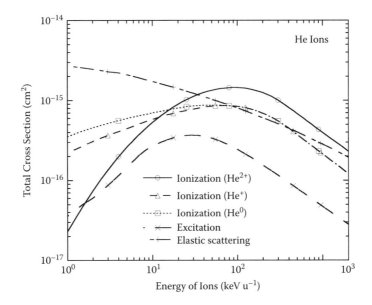

FIGURE 15.3
Total cross sections due to helium ion impact on water vapor.

Total ionization cross sections for He⁺ were fitted to the experimental data of Rudd et al. (1985c) measured between 1 and 100 keVu⁻¹ and Toburen et al. (1980) between 75 and 500 keVu⁻¹. The experimental cross sections for He⁰ ionization are not available. Therefore, TICSs for He⁰ at energies lower than 100 keVu⁻¹ were adjusted to fit the electronic stopping powers tabulated in ICRU Report 49 (1993). In the region below 30 keVu⁻¹ the stopping powers for He²⁺ are negligible and He⁰ becomes the main contributor. This point is elucidated further in Section 15.5. At energies above 100 keVu⁻¹, the total cross sections were assumed to be the same as those of He⁺. Figure 15.3 shows calculated TICSs for this work.

15.2 Excitation

Impact excitation data of all helium ions on water molecules are not available. Therefore, we tentatively use the model of Miller and Green (1973), which was described for protons in Section 14.2. We have assumed cross sections for protons can be applied to the impact excitation cross sections for all three helium ions (He²⁺, He⁺, and He⁰). Although this assumption may seem to be too naïve, the fractional energy deposition due to excitation is at most about 10% for the overall energy range. Therefore, a serious inconsistency does not occur. Figure 15.3 shows the calculated total excitation cross section as a function of helium energy.

15.3 Elastic Scattering

Elastic scattering for α-particles was evaluated using the classical mechanics trajectory calculations (CMTCs), which were described for protons in Section 14.3. For α-particles, some minor changes are needed in parameters used for proton elastic scattering. Ziegler and coworkers (1985) derived the optimum parameters for α-particles ($z = 2$). The screening length r_s is given by

$$r_s = 0.88534 r_B (z^{0.23} + Z^{0.23})^{-1} \tag{15.3}$$

where r_B is the Bohr radius and 0.88534 is a numerical constant from the Thomas-Fermi model of an atom. A screening function $F_s(r/r_s)$ modifying the Coulomb potential for α-particles is given by

$$F_s(r/r_s) = 0.1818e^{-3.2r/r_s} + 0.5099e^{-0.942r/r_s} + 0.2802e^{-0.4029r/r_s} + 0.2817e^{-0.2016r/r_s} \tag{15.4}$$

The calculation procedures are the same as for protons. The minimum impact parameter p_{min} was fixed at 0.01, which corresponds to the backward scattering angle $\sim\pi$. The maximum value p_{max}, which corresponds to the forward scattering, was chosen to reproduce the nuclear stopping powers given in the ICRU database (1993). The p_{max} values increase from 48 to 29,100 for the α-particle energies 1 to 2 MeVu^{-1}. The total elastic scattering cross sections are shown in Figure 15.3. We approximated both total and differential elastic scattering cross sections for He$^+$ and He^0 to be equal to those for He^{2+}.

EXAMPLE 15.1

Write a Monte Carlo program for sampling the scattering angle of an α-particle by elastic scattering and calculate the energy lost by the projectile during the collision.

SOLUTION 15.1

Given that R_i is the random number associated with the scattering angle θ_i, R_i is calculated for N angles ranging from θ_{cut} to π, using

$$R_i(\theta_i) = \int_{\theta_{cut}}^{\theta_i} \frac{d\sigma}{d\Omega} \sin\theta d\theta \left/ \int_{\theta_{cut}}^{\pi} \frac{d\sigma}{d\Omega} \sin\theta d\theta \right.$$

In the sampling procedure for the projectile of energy T and mass m_p, a random number RA is generated and compared with R_i, as follows:

```
RA=rand(seed)
do i=1,N
If(RA ≥ R(i) and RA < R(i+1)) then
```

$$\theta = \frac{\theta_{i+1} - \theta_i}{R_{i+1} - R_i} \cdot (RA - R_i) + \theta_i$$

```
end if
end do
```

$$W(\theta, T) = 4T \frac{m_p M_t}{(m_p + M_t)^2} \sin^2 \frac{\theta}{2} \quad \text{\%the energy lost by the projectile}$$

EXAMPLE 15.2

Compare the maximal energy transfer during elastic scattering for 100 keV proton and α-particle impact on water.

SOLUTION 15.2

$$W_{\max} = 4T \frac{m_p M_t}{(m_p + M_t)^2}$$

For proton impact, the maximal energy transfer is $W_{\max} = 4.100$ keV $\times \frac{1 \cdot 18}{(1+18)^2} = 19.94$ keV, and for α-particle the corresponding value is $W_{\max} = 4 \cdot 100$ keV $\frac{4 \cdot 18}{(4+18)^2} = 59.50$ keV (almost three times larger than for protons of the same energy).

15.4 Charge Transfer

For a complete description of the full slowing down of α-particles we need cross sections for charge transfers. Table 15.1 presents a list of charge transfer interactions for low-energy He²⁺, He⁺, and He⁰ projectiles with water. The present calculation considered six kinds of interactions such as σ_{21} and σ_{20} for He²⁺, σ_{10} and σ_{12} for He⁺, and σ_{01} and σ_{02} for He⁰. Charge transfer can also

TABLE 15.1

Charge Transfer for Low-Energy He²⁺, He⁺, and He⁰ Projectiles with Water

Two-electron capture	$He^{2+} + H_2O \rightarrow He^0 + H_2O^{2+}$	(σ_{20})
One-electron capture	$He^{2+} + H_2O \rightarrow He^+ + H_2O^+$	(σ_{21})
	$He^+ + H_2O \rightarrow He^0 + H_2O^+$	(σ_{10})
One-electron loss	$He^+ + H_2O \rightarrow He^{2+} + H_2O + e$	(σ_{12})
	$He^0 + H_2O \rightarrow He^+ + H_2O + e$	(σ_{01})
Two-electron loss	$He^0 + H_2O \rightarrow He^{2+} + H_2O + 2e$	(σ_{02})
One-electron capture and target ionization	$He^{2+} + H_2O \rightarrow He^+ + H_2O^{2+} + e$	
	$He^+ + H_2O \rightarrow He^0 + H_2O^{2+} + e$	
One-electron loss and target ionization	$He^+ + H_2O \rightarrow He^{2+} + H_2O^+ + 2e$	
	$He^0 + H_2O \rightarrow He^+ + H_2O^+ + 2e$	

be accompanied by simultaneous target ionization for both electron loss and electron capture processes as listed in the table. There is insufficient information on cross sections for these interactions to be included in the calculations.

Experimental cross sections for σ_{21} and σ_{20} for He^{2+} are given by Rudd et al. (1985b) in the energy range 5 to 150 keVu^{-1} in water. These cross sections were least squares fitted by simple polynomial functions of the form $\log \sigma_{if} = \Sigma c_j (\log T)^{j-1}$, where log is to the base 10, and T in keVu^{-1}. We omit individual numerals for parameter c_j for all cases. At energies below 5 keVu^{-1} and above 150 keVu^{-1}, these functions were extrapolated assuming a smooth transition at the boundaries. Cross sections for σ_{10} and σ_{12} for He^+ were fitted to the experimental data of Rudd et al. (1985c) measured between 1 and 100 keVu^{-1} and Sataka et al. (1990) between 75 and 500 keVu^{-1}. In the latter, H_2O data were obtained by the additivity rule from the relationship $\sigma(H_2) + \sigma(O_2)/2$ using the measured values for H_2 and O_2. Extrapolation was made for the energies greater than 500 keVu^{-1}. Electron loss cross section (σ_{01}) for He^0 was least squares fitted using experimental data of Allison (1958) below 100 keVu^{-1} and Sataka et al. (1990) between 75 and 500 keVu^{-1}. In the former, H_2O data were obtained from the relationship $2\sigma(H) + \sigma(O)$ using the measured values for H and O. Cross sections for two-electron loss(σ_{02}) were fitted to the data of Sataka et al. (1990). Smooth extrapolation was carried out where the experimental data were lacking. Figure 15.4 shows the total cross sections of electron capture processes (a) and electron loss processes (b) based on available experimental data.

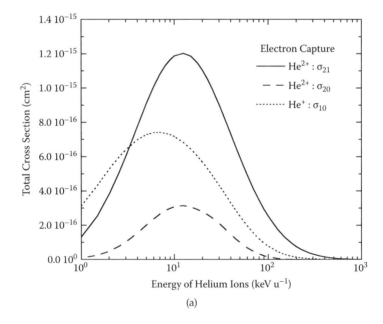

(a)

FIGURE 15.4
(a) Total cross sections of electron capture due to He ion impact on water vapor. (b) Total cross sections of electron loss due to He ion impact on water vapor.

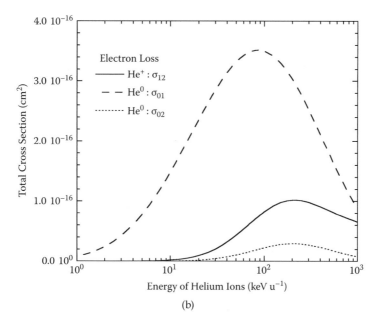

(b)

FIGURE 15.4
(Continued)

15.5 Stopping Powers

15.5.1 Electronic Stopping Powers

The reliability of cross sections appropriate to the macroscopic stopping of low-energy α-particles in water were examined. We performed analytical calculations of electronic stopping powers for low-energy α-particles. The electronic stopping powers for α-particles can be obtained using expressions similar to those for protons (Equations (14.17) and (14.18)). The stopping cross section σ_{stop} is given by

$$\sigma_{stop} = F_{++}(Q_{i++}\sigma_{i++} + Q_e\sigma_e + Q_{20}\sigma_{20} + Q_{21}\sigma_{21})$$

$$+ F_+(Q_{i+}\sigma_{i+} + Q_e\sigma_e + Q_{10}\sigma_{10} + Q_{12}\sigma_{12}) \qquad (15.5)$$

$$+ F_0(Q_{i0}\sigma_{i0} + Q_e\sigma_e + Q_{01}\sigma_{01} + Q_{02}\sigma_{02})$$

in which the subscripts ++, +, and 0 denote He^{2+}, He^+, and He^0, respectively. F denotes equilibrium charge fractions and σ denotes total cross sections. The mean energy transfer for ionization is

$$Q_i = \bar{\varepsilon} + \bar{B}_{H_2O} \qquad (15.6)$$

where $\bar{\varepsilon}$ is the average energy of secondary electrons by ion impact, and $\bar{B}_{H_2O}$ is the average binding energy of a water molecule. The average energies of secondary electrons ejected by He^{2+} were obtained from the experimental data (Toburen et al. 1980; Bolorizadeh and Rudd 1986; Toburen and Wilson 1977). For He^+ impact, Equations (15.1) and (15.2) were used to obtain $\bar{\varepsilon}$. The average energy by neutral helium impact was assumed to be equal to that of He^+. Mean energy transfer Q_{if} values in charge exchanges were modeled as

$$Q_{20} = 2\bar{B}_{H_2O} + 2\frac{T}{\lambda'} - 2\bar{B}_{He}, \qquad Q_{21} = \bar{B}_{H_2O} + \frac{T}{\lambda'} - \bar{B}_{He}$$

$$Q_{10} = \bar{B}_{H_2O} + \frac{T}{\lambda'} - \bar{B}_{He}, \qquad Q_{12} = \bar{B}_{He} \qquad (15.7)$$

$$Q_{01} = \bar{B}_{He}, \qquad Q_{02} = 2\bar{B}_{He}$$

in which T is the kinetic energy of the projectile in eV units and $\lambda' = (m_{He}/m_H)$ $\lambda = 4\lambda = 7{,}344$. $\bar{B}_{He} = 24.59$ eV is the average binding energy of a helium atom. Figure 15.5 shows the mean energy transfers calculated using the above equations.

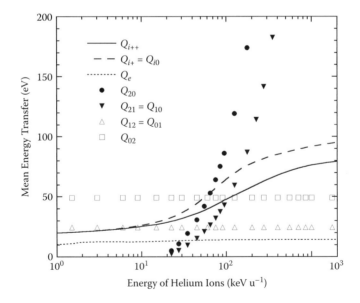

FIGURE 15.5
Mean energy transfers in collisions of helium ions with water vapor leading to emission of secondary electrons or excitations and charge exchange processes.

EXAMPLE 15.3

What is the difference in energy lost by α-particles of the same energy during ionization and double-electron capture?

SOLUTION 15.3

For ionization, the energy lost by the projectile is $Q_i = \varepsilon + \bar{B}_{H_2O}$, while for the double-electron capture process the energy loss is $Q_{20} \approx 2\bar{B}_{H_2O} + 2\frac{T}{\lambda'} - 2\bar{B}_{He}$. Therefore, the difference in energy transferred is $Q_{20} - Q_i \approx \bar{B}_{H_2O} + 2\frac{T}{\lambda'} - 2\bar{B}_{He} - \varepsilon$.

Equilibrium charge fractions, F_{++}, F_+, and F_0, are given by Allison (1958) as follows:

$$F_{++} = [(a-b)\sigma_{20} + g(a+\sigma_{21}) - f(b+\sigma_{21})]/D$$

$$F_+ = (b\sigma_{20} - g\sigma_{21})/D \tag{15.8}$$

$$F_0 = (f\sigma_{21} - a\sigma_{20})/D$$

in which

$$a = -(\beta + \sigma_{21}), \quad b = \sigma_{01} - \sigma_{21}, \quad f = \sigma_{10} - \sigma_{20}, \quad g = -(\alpha + \sigma_{20})$$

$$D = ag - bf, \quad \alpha = \sigma_{01} + \sigma_{02}, \quad \beta = \sigma_{10} + \sigma_{12} \tag{15.9}$$

All parameters can be evaluated using the six types of cross sections for charge exchanges.

The nuclear stopping powers are calculated with the same formula as are protons (Equation (14.26)). Figure 15.6 shows the calculated electronic and nuclear stopping powers in comparison with the published data. Both stopping powers agreed well with the ICRU tabulations (ICRU 1993). Figure 15.7 shows the fractional electronic stopping powers for three helium ions. It is found that dressed helium ions are the dominant component at energy less than ~100 keVu⁻¹.

For the Monte Carlo track structure simulation, as described in Section 14.5, we need to revise the present energy transfer model for the charge exchange interactions. For α-particles, $\Delta E = \bar{B}_{H_2O}$ is applied to the interactions for one-electron capture such as σ_{21} and σ_{10}. We assume the energy deposition for two-electron capture (σ_{20}) becomes $2\bar{B}_{H_2O}$ because H_2O^{2+} ion is generated. In the one-electron loss, the electron generated in the σ_{12} and σ_{01} interactions has the kinetic energy of T/λ, while two electrons are emitted with the same energy and the same direction in the two-electron loss process (σ_{02}). The emission direction of the stripped electron is assumed to be the same as that of the projectile.

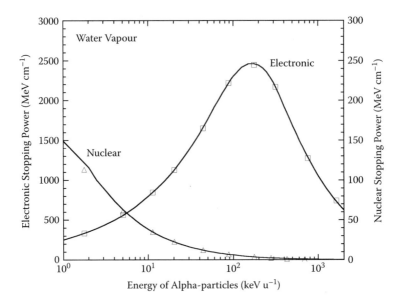

FIGURE 15.6
Comparison of the electronic and nuclear stopping powers of water vapor for α-particles.
Lines: present calculations; symbols: ICRU Report 49 (1993).

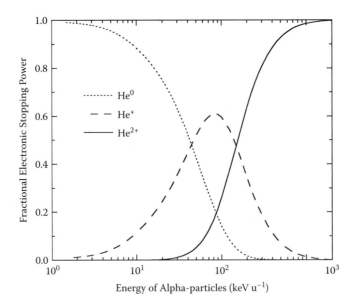

FIGURE 15.7
Fractional electronic stopping powers of water vapor for three helium ions as a function of
energy of α-particles.

EXAMPLE 15.4

What is the energy of an electron emitted by the electron loss process of a 1 MeV/u α-particle?

SOLUTION 15.4

$$\varepsilon(eV) = \frac{T(eV)}{\lambda'} = \frac{T'(eV/u)}{\lambda} = \frac{10^6}{1836} = 544.66$$

15.6 Summary

1. Energy spectra of secondary electrons ejected by He^{2+} impact were calculated using the empirical model of Rudd. Those for He^+ and He^0 were obtained by modifying Rudd's model.

2. Experimental angular distributions of electrons were compiled. Interpolation was made where data are lacking.

3. Total ionization cross sections were obtained by fitting polynomial functions to the experimental data.

4. Excitation cross sections and mean excitation energy loss were treated based on the formula given by Miller and Green.

5. Nuclear elastic scattering was evaluated using the classical mechanics trajectory calculations.

6. Charge transfer cross sections were compiled based on available experimental data.

7. Calculations of the electronic and nuclear stopping power were performed without any parameters, such as the mean excitation energy and the effective charge.

QUESTION

1. Similar to Table 15.1, construct a table of charge transfers for low-energy C^{6+}, C^{5+}, C^{4+}, C^{3+}, C^{2+}, C^{1+}, and C^0 carbon ion projectiles with water.

References

Allison SK. 1958. Experimental results on charge-changing collisions of hydrogen and helium atoms and ions at kinetic energies above 0.2 keV. *Rev. Mod. Phys.* 30: 1137–1168.

Bolorizadeh MA, Rudd ME. 1986. Angular and energy dependence of cross sections for ejection of electrons from water vapor. II. 15–150-keV proton impact. *Phys. Rev.* A33: 888–892.

ICRU. 1993. *Stopping Powers and Ranges for Protons and Alpha Particles.* ICRU Report 49.

ICRU. 1996. *Secondary Electron Spectra from Charged Particle Interactions.* ICRU Report 55.

Miller JH, Green AES. 1973. Proton energy degradation in water vapor. *Radiat. Res.* 54: 343–363.

Rudd ME, Goffe TV, DuBois RD, Toburen LH. 1985a. Cross sections for ionization of water vapor by 7–4000-keV protons. *Phys. Rev.* A31: 492–494.

Rudd ME, Goffe TV, Itoh A. 1985b. Ionization cross sections for 10–300-keV/u and electron-capture cross sections for 5–150-keV/u $^3He^{2+}$ ion in gases. *Phys. Rev.* A32: 2128–2133.

Rudd ME, Itoh A, Goffe TV. 1985c. Cross sections for ionization, capture, and loss for 5–450-keV He^+ on water vapor. *Phys. Rev.* A32:2499–2500.

Sataka M, Yagishita A, Nakai Y. 1990. Measurement of charge-changing cross sections in collisions of He and He^+ with H_2, O_2, CH_4, CO and CO_2. *J. Phys.* B23: 1225–1234.

Toburen LH, Wilson WE. 1977. Energy and angular distributions of electrons ejected from water vapor by 0.3–1.5 MeV protons. *J. Chem. Phys.* 66: 5202–5213.

Toburen LH, Wilson WE, Popowich RJ. 1980. Secondary electron emission from ionization of water vapor by 0.3- to 20-MeV He^+ and He^{2+} ions. *Radiat. Res.* 82: 27–44.

Uehara S, Nikjoo H. 2002. Monte Carlo track structure code for low-energy alpha-particles in water. *J. Phys. Chem.* B106: 11051–11063.

Ziegler JF, Biersack JP, and Littmark U. 1985. *The Stopping and Range of Ions in Solids.* New York: Pergamon Press. http://www.srim.org/, new edition in 2003.

16

Cross Sections for Interactions of High-Energy Protons (>1 MeVu^{-1}) in Water

16.1 Ionization

16.1.1 Secondary Electrons

The collision of a charged particle with another at rest is described by the Rutherford scattering formula (ICRU 1996):

$$\frac{d\sigma_R}{d\varepsilon} = \frac{4\pi a_0^2}{T} \left(\frac{R}{E}\right)^2 \tag{16.1}$$

where ε is the kinetic energy of the secondary electron after the collision, T is the kinetic energy of an electron with the same speed as the incident proton, i.e., $T = T_p/\lambda$, a_0 is the Bohr radius, R is the Rydberg energy (13.6 eV), and $E = \varepsilon + B$ is the energy transfer in which B is the binding energy of a molecular orbital.

Based on the Rutherford formula, the binary encounter approximation treats the collision as a classical one between the projectile and a single electron in the target. A simpler form of the binary encounter theory leads to a singly differential cross section of the form (ICRU 1996)

$$\frac{d\sigma_{BE}}{d\varepsilon} = \left(\frac{d\sigma_R}{d\varepsilon}\right) \left(1 + \frac{4U}{3E}\right) \qquad \text{for } E_{min} \leq E \leq E_- \tag{16.2a}$$

$$\frac{d\sigma_{BE}}{d\varepsilon} = \left(\frac{d\sigma_R}{d\varepsilon}\right) \left(\frac{U}{6E}\right)\left\{\left(\frac{4T}{U}\right)^{3/2} + \left[1 - \sqrt{1 + \frac{E}{U}}\right]^3\right\} \qquad \text{for } E_- \leq E \leq E_+ \tag{16.2b}$$

and

$$\frac{d\sigma_{BE}}{d\varepsilon} = 0 \qquad \text{for } E > E_+ \tag{16.2c}$$

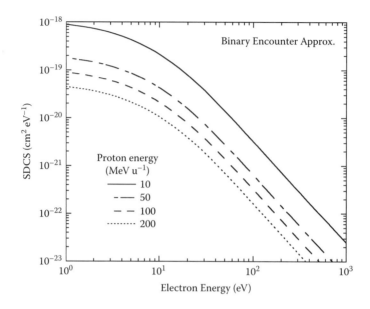

FIGURE 16.1
Energy spectra calculated using the binary encounter approximation for secondary electrons ejected by proton impact.

where U is the kinetic energy of the target electron and

$$E_\pm = 4T \pm 4(TU)^{1/2} \qquad\qquad (16.2d)$$

The calculated spectra were justified by comparing with the available experimental data for the proton energies 1.5, 3, and 4.2 MeVu^{-1} (Nikjoo et al. 2008). Figure 16.1 shows the energy spectra of secondary electrons for the proton energies 10, 50, 100, and 200 MeVu^{-1}.

EXAMPLE 16.1
The kinetic energies of electrons in water are 48.36 eV (1b$_1$), 59.52 eV (3a$_1$), 61.91 eV (1b$_2$), 70.71 eV (2a$_1$), and 1589.5 eV (1a$_1$). Calculate $E_\pm$ for 1 MeV protons.

SOLUTION 16.1

$$E_\pm = 4T \pm 4(TU)^{1/2}$$

where

$$T = 10^6 \text{ eV}/1836 = 544.66 \text{ eV}$$

Table Solution 16.1

Orbital	U (eV)	E_- (eV)	E_+ (eV)
1b1	48.36	1529.47	2827.83
3a1	59.52	1458.45	2898.85
1b2	61.91	1444.13	2913.17
2a1	70.71	1393.66	2963.64
1a1	1589.50	539.7*	5900.46

* The calculated E_- is less than zero, which is not allowed. Therefore, $E_- = E_{min}$ (the binding energy of the K-shell electron of oxygen).

For low-energy proton impact, experimental angular distributions of electrons ejected from water by proton impact were used, while experimental data are unavailable for high-energy protons of >1 MeVu⁻¹. The initial emission angle for secondary electrons was approximated by the kinematical relationship

$$\cos\theta = \sqrt{\frac{\lambda\varepsilon}{4T_p}} \qquad (16.3)$$

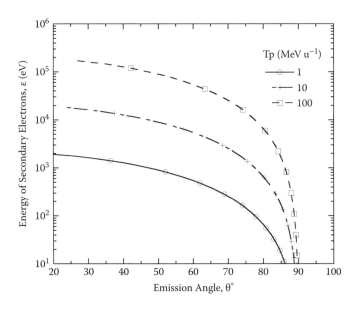

FIGURE 16.2

Kinematical relationship between the emission angle and the energy of secondary electrons, ε, for the proton energy, T_p.

This equation suggests almost all of secondary electrons are ejected vertically from the projectile path. Figure 16.2 shows the emission angle of secondary electrons with the energy ε generated by proton energy T_p. The tracks of secondary electrons were generated by the electron code KURBUC, which is effective for the electron energy of 10 MeV.

EXAMPLE 16.2

What would be the largest angle of the ionized electron within the validity of the kinematical relationship?

SOLUTION 16.2

$$0 \leq \cos\theta = \sqrt{\frac{\lambda\varepsilon}{4T_p}} \leq 1$$

Therefore,

$$0° \leq \theta \leq 90°$$

16.1.2 Total Cross Sections

Total ionization cross sections for high-energy protons were obtained by energy scaling of the electron ionization cross sections. The relationship between the kinetic energy of a proton T_p and that of an electron T_e with the same speed v is given by

$$T_p = \frac{1}{2}m_p v^2 = \lambda T_e \tag{16.4}$$

For example, the electron kinetic energy of 109 keV corresponds to the proton kinetic energy of 200 MeV. The proton cross section at T_p was obtained by such scaling of the electron cross section at T_e. Therefore, the calculation of electron ionization cross sections was carried out using Seltzer's (1988) formula. The detailed descriptions for this theory and various parameters requisite for numerical calculations have already been given in Section 13.1.

Partial and total ionization cross sections were calculated from 10 eV to the mega-electron-volt region for electron energies. Both the cross sections for the energy range of electrons between 545 eV and 109 keV were used for cross sections for the corresponding proton energy range between 1 and 200 MeV. A minor adjustment was made for the energies higher than 150 MeV in order to reproduce the proper continuous slowing down ranges and the stopping powers. Figure 16.3 shows the total ionization cross sections, including those for energies lower than 1 MeV. The total ionization

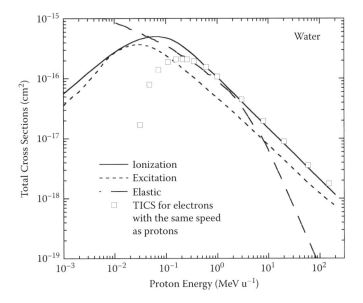

FIGURE 16.3

Total cross sections due to proton impact on water. The ionization cross section is compared with that for electrons with the same speed as protons.

cross sections (TICSs) for electrons with the same speed as protons are plotted for a comparison.

16.2 Excitation

Equation (14.5), given by Miller and Green (1973), was used for the estimation of excitation cross sections and mean excitation energy loss by proton impact in the energy range greater than 1 MeVu^{-1}. Figure 16.3 shows the total excitation cross sections, including those for energies lower than 1 MeVu^{-1}.

16.3 Elastic Scattering

Total elastic cross sections for high-energy protons were obtained by the classical mechanics trajectory calculations (CMTCs). This method is just the same as for low-energy protons lower than 1 MeVu^{-1} (Section 14.3). The maximum impact parameter, p_{max}, which determines the behavior at the small angle, was chosen to reproduce the published nuclear stopping powers. The p_{max} values converge 13,500 for the energy greater than 25 MeVu^{-1}. Figure 16.3 shows the total elastic cross sections, including those for energies lower than 1 MeVu^{-1}.

In the early stages of the development of data compilation, angular distributions of elastic scattering calculated by the CMTC were examined. The Monte Carlo full slowing down tracks showed unrealistic tortuous tracks for the energies greater than 20 MeV. Therefore, we adopted an alternative formula called the Mott scattering formula, effective for high- energy protons (ICRU 1978):

$$\frac{d\sigma_{el}(\theta)}{d\Omega} = \frac{1}{4} N_A \frac{Z^2}{A} z^2 r_e^2 \left(\frac{m_e c}{P\beta}\right)^2 \frac{1 - \beta^2 \sin^2\left(\frac{\theta}{2}\right)}{\sin^4\left(\frac{\theta}{2}\right)} \tag{16.5}$$

where N_A = the Avogadro constant and A = the molar mass of the scattering medium. The cutoff angles, θ_{cut}, were given by Equation (14.14). The Mott formula was derived assuming the repulsive Coulomb force, not taking into account the screening by atomic electrons for the relativistic energy of projectiles. At the nonrelativistic limit, this formula is reduced to the classical Rutherford scattering formula. Both formulas provide less deflective tracks than the CMTC results because the screening effect is neglected.

16.4 Summary

1. Energy spectra of secondary electrons ejected by high-energy proton impact were calculated using the binary encounter approximation.

2. The initial emission angle for secondary electrons was approximated by the kinematical relationship.

3. Total ionization cross sections were obtained by energy scaling of the electron ionization cross sections; those are evaluated by Seltzer's formula.

4. Excitation cross sections and mean excitation energy loss were treated based on the formula given by Miller and Green.

5. Total cross sections for elastic scattering were evaluated using the classical mechanics trajectory calculations, while the Mott scattering formula was effective for angular distributions of high-energy protons.

QUESTIONS

1. It is usual to use the Fano plot for test of reliability and consistency of experimental cross sections. Describe how this is done.
2. The Born cross section for ionization by electrons can be expressed as $\sigma_B = 4\pi a_0^2/(T/R) [A\ln(T/R) + B]$, where A and B are Bethe constants. Write expressions for A related to the oscillator strength.

References

ICRU. 1978. *Basic Aspects of High Energy Particle Interactions and Radiation Dosimetry.* ICRU Report 28.

ICRU. 1996. *Secondary Electron Spectra from Charged Particle Interactions.* ICRU Report 55.

Miller JH, Green AES. 1973. Proton energy degradation in water vapor. *Radiat. Res.* 54: 343–363.

Nikjoo H, Uehara S, Emfietzoglou D, Brahme A. 2008. Heavy charged particles in radiation biology and biophysics. *New J. Phys.* 10: 075006. http://www.njp.org/.

Seltzer SM. 1988. Cross sections for bremsstrahlung production and electron-impact ionization. In *Monte Carlo Transport of Electrons and Photons*, ed. TM Jenkins, WR Nelson, A Rindi. New York: Plenum, pp. 81–114.

17

Model Calculations Using Track Structure Data of Electrons

17.1 Ranges and W Values

The electron track code KURBUC provides all coordinates of molecular interaction in water vapor, the amount of energy deposited at each event, and the type of interaction at each event. The code simulates the full slowing down of electrons, following the primary electron down to cutoff energy of 7.4 eV, at which point the residual energy is deposited at a random short distance. Tracks were analyzed to provide confirmation on the reliability of the code and information on physical quantities, such as range, W values, dose profiles, and various microdosimetric parameters derived from model calculations. The calculated ranges in water vapor (density $= 1$ g cm^{-3}) shown in Figure 17.1 agree with the continuous slowing down approximation (CSDA) ranges of ICRU (1984) within less than 10% deviation for the starting energy of electrons between 100 eV and 1 MeV. The W values are derived from scoring the number of electron-H$_2$O$^+$ pairs due to the ionization process. Table 17.1 shows the calculated W values in water vapor in comparison with experimental data as a function of the starting energy of electrons (Nikjoo et al. 2006). The calculated W value agreed with the experimental data (Combecher 1980).

17.2 Depth-Dose Distributions

Depth-dose distribution is a severe criterion on the track structure code, because it is affected by not only stopping powers but also deflections. We have examined a capability of the event-by-event electron track structure code KURBUC to reproduce the macroscopic dosimetric parameters such as electron depth-dose distributions over the mm ranges. For electrons with energies above 10 keV, many Monte Carlo transport codes use the

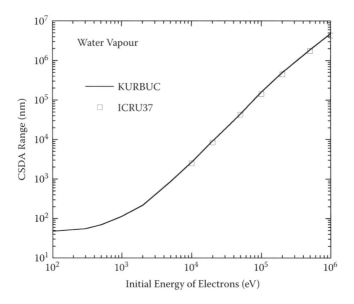

FIGURE 17.1
Calculated CSDA ranges as a function of the energy of electrons in comparison with the data of ICRU Report 37 (1984).

condensed history technique, which approximates the electron degradation by multiple scattering theories in combination with a small number of catastrophic events and uses the restricted stopping power due to ionization and excitation processes to describe the slowing down of electrons. Therefore, individual interaction events do not appear in the calculated track. On the other hand, a major advantage of such codes is highly efficient in time and cost.

In order to examine the performance of different treatments for radiation histories, we have undertaken a benchmark test on electron depth-dose calculations. Absolute dose per unit incident fluence as a function of depth in water of the density $= 1$ g cm^{-3} was calculated for a broad parallel beam of electrons that start at a position within an infinite water phantom. Figure 17.2 shows a comparison of depth-dose curves between KURBUC and CHMC (condensed history Monte Carlo) code MCEP (Uehara 1986), for electron energies 50 keV (a) and 200 keV (b). The distributions obtained by the code KURBUC are in good agreement with MCEP, including the critical characteristics of back scattering contributions. It is concluded that consistent results can be obtained from both the microscopic and macroscopic codes for depth-dose of electrons with energies up to several hundred keV. Such intercomparisons and also with other physical quantities have been useful means for checking the reliability of the codes.

TABLE 17.1

Calculated W Values in Water Vapor as a Function of the
Starting Energy of Electrons (Nikjoo et al. 2006) in
Comparison with the Experimental Data (Combecher 1980)

Electron Energy (eV)	Calculations	Combecher
15.6	127.87	118.65
16.6	121.17	99.17
18.7	70.57	79.59
22.7	62.88	62.74
24.7	55.38	58.32
26.7	55.16	55.47
28.7	50.00	52.13
30.7	49.12	50.13
34.7	45.48	47.00
40.7	42.35	44.17
45.7	40.37	42.06
50.7	41.45	40.63
60.7	38.54	39.04
70.7	37.61	37.46
80.7	36.03	36.67
100.7	35.61	35.15
125.7	34.50	34.13
150.7	33.73	33.89
200.7	32.70	32.70
250.7	32.38	32.40
300.7	32.45	32.03
350.7	31.83	31.75
400.7	31.74	31.29
500.0	31.52	30.89
600.0	31.61	30.53
800.0	30.96	30.61
1,000.0	30.91	
3,000.0	30.53	
10,000.0	30.52	

17.3 Electron Slowing Down Spectra

The electron slowing down spectra are those of the actual energies of secondary electrons at any point in the irradiated material. Those are equivalent to the fluence distribution, differential in energy, of both primary and all subsequent generations of secondary electrons generated throughout the

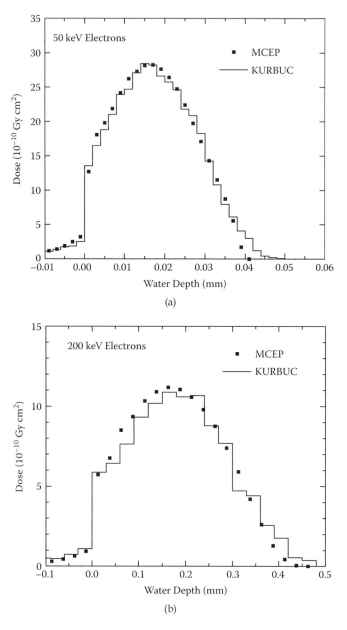

FIGURE 17.2
Comparison of calculated depth-dose curve for a broad parallel beam of electrons with 50 keV
(a) and 200 keV (b). The starting position of the beam is at the origin of depth within an infinite
water phantom.

slowing down process. The definition of the slowing down spectrum $y(E, E_0)$ is as follows (ICRU 1983):

$$y(E, E_0) = \frac{\int_E^{E_0} n(E')dE'}{S(E)} \tag{17.1}$$

where E_0 is the maximum energy of incident electrons, $n(E')$ is the electron spectrum produced in water (number of particles set in motion per unit of energy and volume), and $S(E)$ is the linear collision stopping power at the energy, E. The electron track structure code KURBUC can provide the spectrum $n(E)$, which is obtained by accumulating the absolute number of all secondary and higher-order electrons created by complete slowing down of incident electrons. The secondary electron fluence, differential in electron energy per unit dose, is given in units of cm^{-2} eV^{-1} Gy^{-1} by

$$\frac{\Phi_E}{D} = y(E, E_0)\frac{\rho}{\overline{E_0}} \tag{17.2}$$

where D represents unit dose, ρ is the density of the target material, and $\overline{E_0}$ is the mean initial energy of the electrons (Tilly et al. 2002). Consequently, for tritium having a continuous β-ray spectrum, $\overline{E_0}$ becomes 6.2 keV. For ^{60}Co photons, protons, and α-particles, the $\overline{E_0}$ values are the mean energies of electrons set in motion with a spectrum of their initial energies. The relevant energy calculated from the first-generation electron spectrum produced by ^{60}Co photons is 591 keV (Uehara and Nikjoo 1996). The energy spectra of electrons ejected by ion impact were calculated by the KURBUC ion codes for low-energy protons and α-particles (Uehara et al. 2001; Uehara and Nikjoo 2002). In this case these electrons are used as the input data for the KURBUC electron code, which calculates slowing down spectra. The mean energy of ejected electrons, $\overline{E_0}$, is 56 eV for 0.5 MeVu^{-1} protons and 67 eV for 1 MeVu^{-1} α-particles, respectively. A comparison of the slowing down spectra for such projectiles with the calculations by Tilly et al. and Paretzke (1987) has been reported already (Nikjoo et al. 2006). The validity of the present calculations was examined by comparison with the data of them except for a high-energy electron such as 1 MeV and ^{60}Co. This discrepancy may come from the difference of the calculation model for a secondary electron spectrum between them. Figure 17.3a shows the calculated electron slowing down spectra in water vapor for electrons with initial energies of 5, 50, and 500 keV and 5 MeV electrons. Figure 17.3b shows the calculated electron slowing down spectra in water vapor for protons and α-particles with the energies 0.2 MeVu^{-1} and 2 MeVu^{-1}.

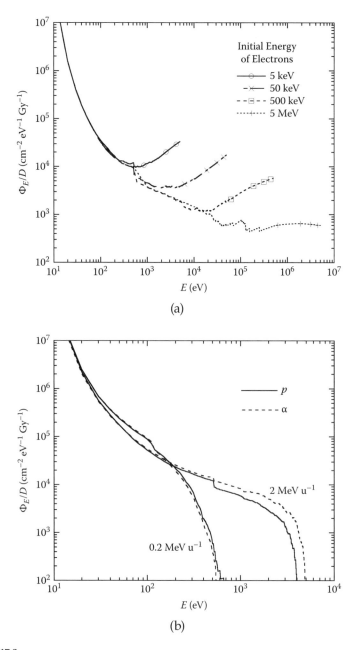

FIGURE 17.3
(a) Slowing down spectra in water vapor for electrons with initial energies of 5, 50, and 500 keV and 5 MeV. (b) Slowing down spectra in water vapor for protons and α-particles with initial energies of 0.2 and 2 MeVu⁻¹.

17.4 Summary

1. The electron track code KURBUC provides all coordinates of molecular interaction in water vapor, the amount of energy deposited at each event, and the type of interaction at each event.

2. The calculated CSDA ranges and the W values in water vapor agreed with the experimental data within less than 10% deviation.

3. Depth-dose curves obtained by the code KURBUC were in good agreement with the results of a CHMC code, including the critical characteristics of back scattering contributions.

4. The electron slowing own spectra in water vapor were obtained for electrons, γ-rays, protons, and α-particles with various initial energies.

References

Combecher D. 1980. Measurements of W values of low-energy electrons in several gases. *Radiat. Res.* 84: 189–218.

ICRU. 1983. *Microdosimetry*. ICRU Report 36.

ICRU. 1984. *Stopping Powers for Electrons and Positrons*. ICRU Report 37.

Nikjoo H, Uehara S, Emfietzoglou D, Cucinotta FA. 2006. Track-structure codes in radiation research. *Radiat. Measure.* 41: 1052–1074.

Paretzke HG. 1987. Radiation track structure theory. In *Kinetics of Nonhomogeneous Processes*, ed. GR Freeman. New York: Wiley, pp. 89–170.

Tilly N, Fernandez-Varea JM, Grusell E, Brahme A. 2002. Comparison of Monte Carlo calculated electron slowing-down spectra generated by ^{60}Co γ-rays, electrons, protons and light ions. *Phys. Med. Biol.* 47: 1303–1319.

Uehara S. 1986. The development of a Monte Carlo code simulating electron-photon showers and its evaluation by various transport benchmarks. *Nucl. Instrum. Meth.* B14: 559–570.

Uehara S, Nikjoo H. 1996. Energy spectra of secondary electrons in water vapour. *Radiat. Environ. Biophys.* 35: 153–157.

Uehara S, Nikjoo H. 2002. Monte Carlo track structure code for low-energy alpha-particles in water. *J. Phys. Chem.* B106: 11051–11063.

Uehara S, Toburen LH, Nikjoo H. 2001. Development of a Monte Carlo track structure code for low-energy protons in water. *Int. J. Radiat. Biol.* 77: 139–154.

18

Model Calculations Using Track Structure Data of Ions

18.1 KURBUC Code System for Heavy Particles

The KURBUC code system is a suite of Monte Carlo track structure codes for the simulation of tracks of full slowing down tracks of ions and electrons in water. The low-energy proton code *Kurbuc_proton* is effective in the energy range between 1 keVu^{-1} and 1 MeVu^{-1} (Uehara et al. 2001). The proton code has been extended to the high-energy range up to 200 MeVu^{-1} (Nikjoo et al. 2008). Recently, Liamsuwan et al. (2011) improved this code to cover up to 300 MeVu^{-1}. The low-energy α-particle code *Kurbuc_alpha* encompasses the energy range between 1 keVu^{-1} and 2 MeV u^{-1} (Uehara and Nikjoo 2002). The tracks of secondary electrons were generated using the electron track code *Kurbuc_electron* (Uehara et al. 1993). The codes can be operated to generate tracks in both the track segment and the full slowing down modes. Figure 18.1 shows (a) the flowchart of the main routine for the proton code, and (b) that of the subroutine for the secondary electrons.

18.2 Ranges and *W* Values

The ion codes simulate the full slowing down of ions, following the primary ion down to cutoff energy of 1 keV, at which point the residual energy is deposited at a random short distance. Tracks were analyzed to provide confirmation on the reliability of the code and information on physical quantities, such as range, *W* values, dose profiles, and various microdosimetry parameters derived from model calculations. The calculated ranges in water vapor (density = 1 g cm^{-3}) shown in Figure 18.2 agree with the continuous slowing down approximation (CSDA) ranges of ICRU (1993) within a few percent deviation for the starting energy of protons between 1 keVu^{-1} and 200 MeVu^{-1} and α-particles between 1 keVu^{-1} and 2 MeVu^{-1}. Table 18.1 shows

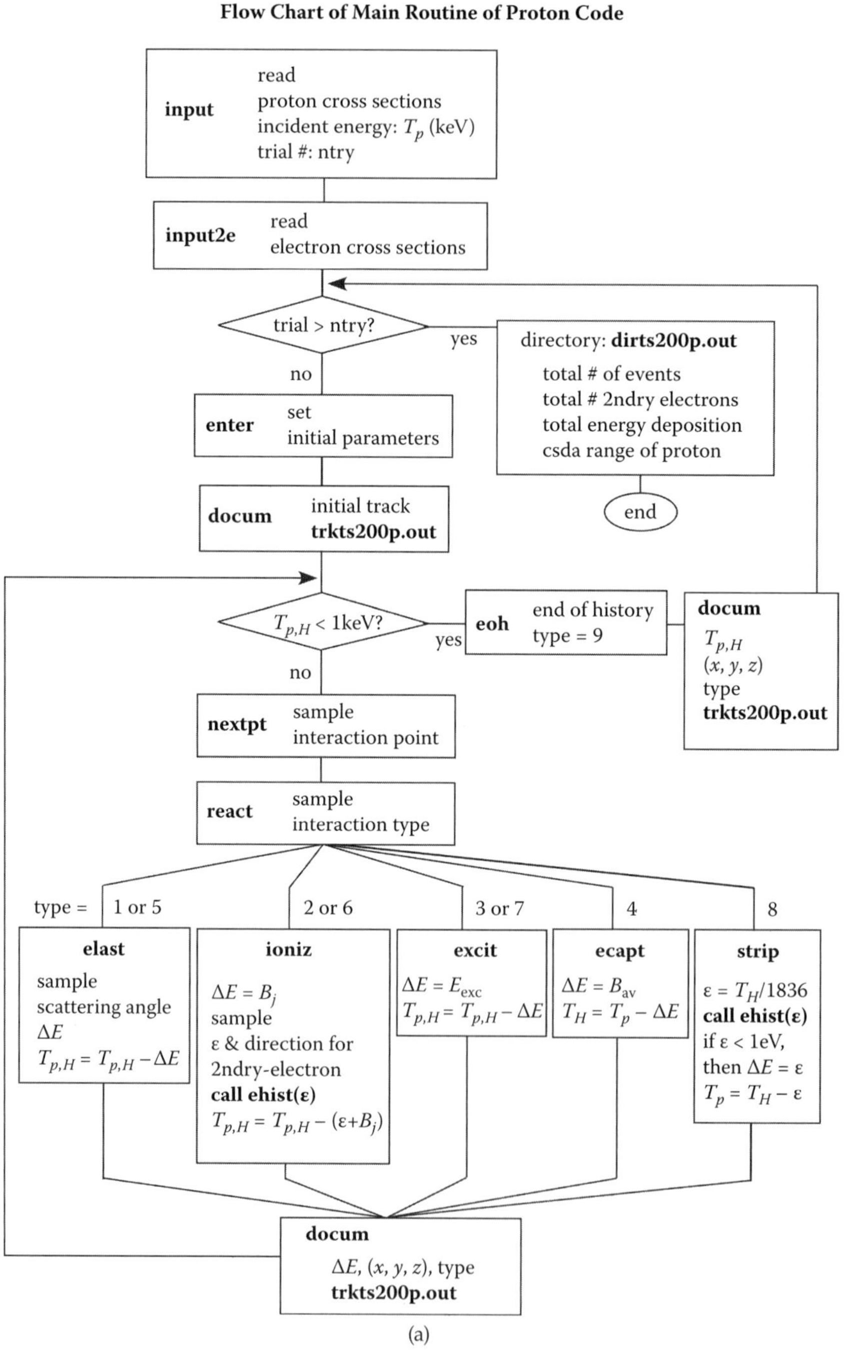

Flow Chart of Main Routine of Proton Code

input — read proton cross sections, incident energy: T_p (keV), trial #: ntry

input2e — read electron cross sections

trial > ntry? — yes → directory: **dirts200p.out**, total # of events, total # 2ndry electrons, total energy deposition, csda range of proton → end

no

enter — set initial parameters

docum — initial track **trkts200p.out**

$T_{p,H} < 1\,\mathrm{keV}$? — yes → **eoh** end of history type = 9 → **docum** $T_{p,H}$, (x, y, z), type **trkts200p.out**

no

nextpt — sample interaction point

react — sample interaction type

type = 1 or 5 | 2 or 6 | 3 or 7 | 4 | 8

elast
sample scattering angle
ΔE
$T_{p,H} = T_{p,H} - \Delta E$

ioniz
$\Delta E = B_j$
sample ε & direction for 2ndry-electron
call ehist(ε)
$T_{p,H} = T_{p,H} - (\varepsilon + B_j)$

excit
$\Delta E = E_{\mathrm{exc}}$
$T_{p,H} = T_{p,H} - \Delta E$

ecapt
$\Delta E = B_{\mathrm{av}}$
$T_H = T_p - \Delta E$

strip
$\varepsilon = T_H/1836$
call ehist(ε)
if $\varepsilon < 1\mathrm{eV}$,
then $\Delta E = \varepsilon$
$T_p = T_H - \varepsilon$

docum
ΔE, (x, y, z), type
trkts200p.out

(a)

FIGURE 18.1
Flowchart of the main routine for the proton code (a) and that of the subroutine for the secondary electrons (b).

Flow Chart of Subroutine Ehist(ε)

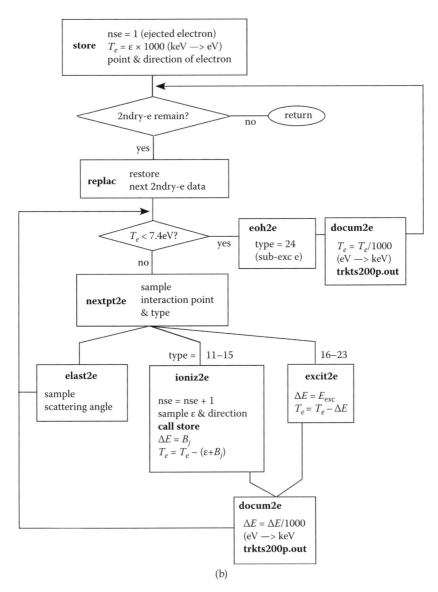

(b)

FIGURE 18.1
(Continued)

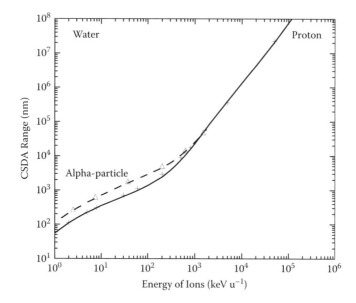

FIGURE 18.2
Calculated CSDA ranges as a function of the energy of protons and α-particles in comparison with the data of ICRU Report 49 (1993).

TABLE 18.1

Calculated *W* Values in Water Vapor as a Function of the Starting Energy of Protons and α-Particles (Nikjoo et al. 2006)

Proton Energy (keVu⁻¹)	*W* (eV)	α-Particle Energy (keVu⁻¹)	*W* (eV)
1.0	52.38	1.0	71.57
1.5	45.94	1.25	63.92
2.0	41.17	1.5	56.24
2.5	37.78	1.75	53.19
3.0	35.72	2.0	50.37
4.0	33.47	2.25	48.44
5.0	32.13	2.5	47.60
6.0	30.28	3.0	42.03
7.0	29.79	4.0	38.88
8.0	29.13	5.0	37.72
9.0	28.83	6.0	35.14
10.0	28.60	7.0	34.54
12.5	28.34	8.0	33.90
15.0	28.46	10.0	33.78
17.5	28.32	15.0	32.86

(*continued*)

TABLE 18.1 (CONTINUED)

Calculated W Values in Water Vapor as a Function of the Starting Energy of Protons and α-Particles (Nikjoo et al. 2006)

Proton Energy (keVu^{-1})	W (eV)	α-Particle Energy (keVu^{-1})	W (eV)
20.0	28.42	20.0	32.46
25.0	28.45	40.0	32.12
30.0	28.94	50.0	32.20
40.0	29.16	100.0	32.28
50.0	29.22	200.0	32.14
60.0	29.43	500.0	32.28
70.0	29.43	1,000.0	32.25
80.0	29.62	1,500.0	32.20
100.0	29.85		
150.0	30.03		
200.0	30.22		
300.0	30.48		
400.0	30.32		
500.0	30.51		
600.0	30.69		
800.0	30.35		
1,000.0	30.53		

the calculated W values in water vapor as a function of the starting energy of ions. The calculated W value at the high-energy limit agreed with the experimentally determined asymptotic limits.

18.3 Depth-Dose Distributions

Depth-dose distribution is a severe criterion on the track structure code, because it is affected by not only stopping powers but also ion deflections. Figure 18.3 shows the depth-dose curve in water phantom for a broad parallel beam of protons with 200 MeV. The depth of the Bragg peak is in agreement with the CSDA range of 26 cm for liquid water. This supports an adoption of the Mott scattering formula for angular distributions for high-energy protons. The calculated distribution implies the capability of the microscopic track structure to obtain such a macroscopic quantity in the ~cm region. The contribution of secondary electrons to the total dose amounts to 70% over the whole depth. Figure 18.4 shows the depth-dose curve in water for α particles with 5 MeV.

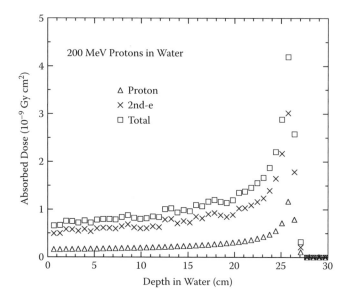

FIGURE 18.3
Calculated depth-dose curves for protons and secondary electrons and their sum for a broad parallel beam of 200 MeV protons.

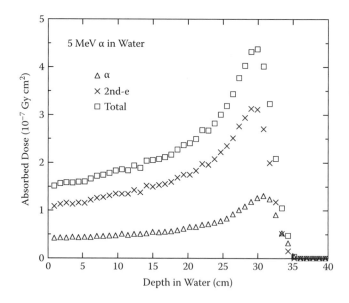

FIGURE 18.4
Calculated depth-dose curves for α-particles and secondary electrons and their sum for a broad parallel beam of 5 MeV α-particles.

18.4 Radial Dose Distributions

The radial dose distribution, which is the energy absorbed locally at a certain radial distance from the path of a primary particle, has been used to check the reliability of track structure calculations. Figure 18.5 shows the radial dose profiles around the path of ion tracks with various energies of protons. Figure 18.6 shows those for α-particles. According to the definition of radial dose distribution, the depositions along the ion track itself are excluded in dose calculations. Energies deposited only due to secondary electrons were accumulated within a coaxial shell with a radial interval of 1 nm along the ion track with a short length. Reasonable agreement is observed over a wide range between the present work and various published data.

18.5 Restricted Stopping Powers

Radius restricted stopping power, L_r, is defined as that part of the total energy loss dE/dl that is deposited within a cylinder of radius r and length dl, centered along the particle track (ICRU 1970). These are useful quantities for checking the reliability of track structure calculations. Figure 18.7 shows

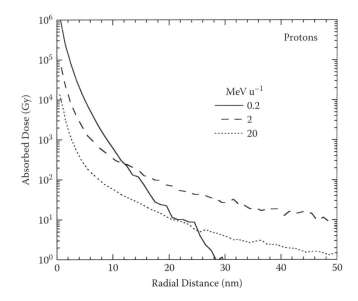

FIGURE 18.5
Calculated radial dose profiles for proton paths with the energies 0.2, 2, and 20 MeVu^{-1}.

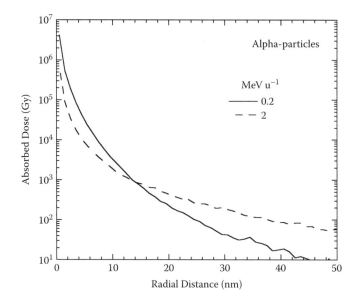

FIGURE 18.6
Calculated radial dose profiles for α-particles paths with the energies 0.2 and 2 MeVu⁻¹.

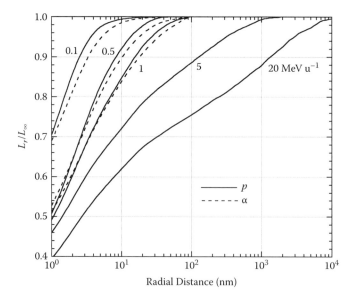

FIGURE 18.7
Ratios of the radius restricted stopping power to the unrestricted stopping power for the ion energies of 0.1, 0.5, and 1 MeVu⁻¹ and for the proton energies 5 and 20 MeVu⁻¹ as a function of radial distance from the ion path in water.

the calculated ratio of L_r to L_∞ for ion energies of 0.1, 0.5, and 1 MeVu^{-1} and for the proton energies 5 and 20 MeVu^{-1} as a function of radial distance from the ion path. A large fraction of the energy lost by fast ions stays within the first nanometer around the path, but with increasing ion energy a large fraction can be transported by energetic electrons to large distances from the path.

18.6 Summary

1. The track structure codes for protons and α-particles enable generation of the full slowing down tracks, including the tracks of all secondary electrons.

2. The calculated CSDA ranges and the *W* values in water vapor agreed with the experimental data within less than 10% deviation.

3. Depth-dose curves calculated by the proton code were in good agreement with the experimental data.

4. The radial dose distributions and the radius restricted stopping powers were derived from the model calculations using the generated track data.

References

ICRU. 1970. *Linear Energy Transfer*. ICRU Report 16.
ICRU. 1993. *Stopping Powers and Ranges for Protons and Alpha Particles*. ICRU Report 49.
Liamsuwan T, Uehara S, Emfietzoglou D, Nikjoo H. 2011. Physical and biophysical properties of proton tracks of energies 1 keV to 300 MeV in water. *Int. J. Radiat. Biol.* 87: 141–160.
Nikjoo H, Uehara S, Emfietzoglou D, Brahme A. 2008. Heavy charged particles in radiation biology and biophysics. *New J. Phys.* 10: 075006. http://www.njp.org/.
Nikjoo H, Uehara S, Emfietzoglou D, Cucinotta FA. 2006. Track-structure codes in radiation research. *Radiat. Measure.* 41: 1052–1074.
Uehara S, Nikjoo H. 2002. Monte Carlo track structure code for low-energy alpha-particles in water. *J. Phys. Chem.* B106: 11051–11063.
Uehara S, Nikjoo H, Goodhead D.T (1993) Cross-sections for water vapour for Monte Carlo electron track structure code from 10eV to 10MeV region. *Physics in Medicine and Biology* 38: 1841–1858
Uehara S, Toburen LH, Nikjoo H. 2001. Development of a Monte Carlo track structure code for low-energy protons in water. *Int. J. Radiat. Biol* 77: 139–154.

Section III

19

Inelastic Scattering of Charged Particles in Condensed Media: A Dielectric Theory Perspective

19.1 Introduction

The inelastic scattering of charged particles by atomic electrons leads to electronic excitations (including ionizations) in the medium. This so-called *electronic energy loss* is the dominant slowing down mechanism for electrons at practically *all* energies above the band-gap (or Fermi energy for metals) and for protons or other light ions above ~100 keV/amu.

The need for studying the inelastic scattering of charged particles in *condensed media* stems from the fact that for most applications in radiation science the irradiated materials are in the condensed phase; that is, they are either liquids or solids. The electronic energy loss of charged particles in condensed media exhibits important differences from the gas phase owing to the long-range polarization of the medium by the charged particle field. As a result of this polarization, the interaction of the charged particle with the target electrons in condensed media is mediated by a *screened* rather than a bare Coulomb force. The screened interaction has two consequences: it modifies the spectrum of *single-electron* excitations, which, in the condensed phase, take the form of electron-hole pairs,[*] and it gives rise to *many-electron* excitations[†] delocalized over length scales well exceeding atomic dimensions. These collective excitations, which have energy quanta (the so-called plasmons) at about 20 eV in biological media, involve a large number of weakly bound valence or conduction electrons and have no counterpart in the gas phase. In contrast, single-electron excitations from inner shells with typical binding energies above ~100 eV are well localized to single atoms and, to a good approximation, remain the same in both the gas and condensed phases.

---

[*] To the extent the motion of electrons and holes is *not* correlated, one speaks, in analogy to the gas phase, of an ionization event; this is, for example, always the case in the excitations of inner shell electrons.

[†] These are generally understood as collective oscillations of the electron density.

Contrary to the gas phase, direct experimental measurements of inelastic scattering cross sections in condensed media are impractical (this is especially true for liquid targets). The reason is that due to the close proximity of atoms in the condensed phase, scattering experiments with charged particle beams nearly always measure the outcome of *multiple scattering* events within the target; thus, deducing information about *single scattering* cross sections is not straightforward. Therefore, it is usually said that theory plays a more important role in the condensed than in the gas phase.

The literature on the theory of charged particle-solid interaction is vast. Here we limit our discussion to those aspects of the theory that underline most inelastic scattering calculations in radiation dosimetry and radiation effect studies (in both radiation biophysics and materials science) as well as those that form the basis of many Monte Carlo codes that simulate charged particle transport in condensed media.

In contrast to the low-energy domain, the inelastic scattering of high-energy charged particles with velocities exceeding those of target electrons seems now to be well understood. At the high-energy domain, the Born and Bethe theories furnish a suitable theoretical framework where reasonably accurate results can be obtained over a wide range of particle energies covering most applications of interest. A key role is played by the *dielectric response function* of the material, which accounts in an effective and compact manner for condensed phase effects, such as the presence of collective (plasmon-like) excitations. The dielectric description of inelastic scattering was first introduced by Fermi (1940) in his (classical) treatment of the density effect in stopping power of relativistic particles in condensed media. The quantum mechanical description of inelastic scattering in condensed media was advanced in the 1950s, somewhat independently, by Bohm and Pines (1953), Lindhard (1954), Hubbard (1955), Fano (1956), and Ritchie (1959). The extension of the Bethe theory, originally developed for isolated atoms and molecules, to condensed media using the dielectric theory was made by Fano (1963). An insightful and comprehensive review of the Bethe theory, which, in many respects, is also relevant to inelastic scattering in the condensed phase, has been given by Inokuti (1971). In addition to the above papers, the books by Schattschneider (1986) and Egerton (1996) as well as the monograph of Kaplan and Miterev (1987) are particularly relevant in the present context. The books by Schattschneider and Egerton are at a somewhat more advanced level and from the point of view of electron energy loss spectroscopy. Schattschneider's book emphasizes theoretical aspects through an in-depth discussion of the dielectric formalism, whereas Egerton's book is more relevant to the experimental aspects and their theoretical interpretation; both books are limited to electrons. The monograph of Kaplan and Miterev is toward understanding radiation effects in biological media, and it discusses both electrons and heavy charged particles. Finally, a concise quantum mechanical description of formal scattering theory is presented as a self-contained chapter in the book of Rossi and Zaider (1996).

Several assumptions underline the present treatment. First, we assume that the target is *amorphous* and *bulk*; that is, for all practical purposes its properties do not depend on the orientation and can be considered infinite in all three directions. The latter assumption implies that boundary (or surface) effects in the dielectric response of the target that will influence inelastic scattering are neglected. In general, both assumptions are in line with the majority of Monte Carlo simulations performed in radiation dosimetry, and in particular, they should be reasonable approximations for liquid water targets with small surface-to-volume ratio. Energy losses to surface excitations are known to differ from those in the bulk, most notably due to *surface* plasmons that are red-shifted compared to *bulk* plasmons, as first predicted by Ritchie (1957). However, surface excitations can occur only in the near surface region, which generally extends only up to a few nanometers inside the material (Pauly and Tougaard 2009). Moreover, the reduction of the bulk inelastic scattering probability within the near surface region (the so-called Begrenzungs effect) is, to a good approximation, compensated by the increase of the probability of surface excitations. Thus, except perhaps for nanostructures, surface effects can be safely neglected and a spatially nonvarying bulk inelastic scattering probability can be assumed within the medium (Feibelman 1973). We also assume *incoherent* scattering; that is, effects (if any) arising from the interference of scattered waves from different atoms (or molecules) are neglected. Although it is not clear whether interference effects should be of any concern at all in inelastic collisions that, due to energy loss, cause a change of the particle's wavelength, the incoherent scattering approximation is strictly valid for homogeneous, isotropic, and amorphous materials and at not too low particle energies where the de Broglie wavelength of the particle is larger than the separation distances of the individual scattering sites. Assuming typical interatomic distances in condensed media of the order of ~1 Å, the incoherent scattering approximation seems appropriate for electrons with energies above several tens of eV and for heavy charged particles of practically any kinetic energy.[*] However, the recent work of Liljequist (2008) seems to indicate that even for elastic electron scattering where interference effects are well established (e.g., diffraction effects), the incoherent approximation holds reasonably well down to much lower energies (~10 eV). Finally, we assume nonrelativistic charged particles. This amounts to considering only the *longitudinal* part of the interaction (i.e., forces acting along the momentum vector) while ignoring the transverse part. This choice is made partly because differences between the gas and condensed phase are most pronounced at low particle energies and partly because a simple kinematic correction to the longitudinal expressions will generally suffice for most applications with low atomic number materials (e.g., biological targets) up to moderate relativistic velocities, e.g., for electrons up to ~1 MeV. For a fully relativistic treatment of

---

[*] The de Broglie wavelength of an electron (or positron) of kinetic energy T is $\lambda(\text{Å}) \approx 12/\sqrt{T(\text{eV})}$, whereas for ions of specific kinetic energy T_{ion} is $\lambda(\text{Å}) \sim 0.009/\sqrt{T_{ion}(\text{keV/amu})}$.

the inelastic scattering of charged particles in condensed media relevant to the present context one may consult Fano (1963) or the more recent paper by Fernandez-Varea et al. (2005).

Finally, we should note that many of the calculations that we are concerned with here are greatly simplified by the use of atomic units (a.u.) where $m = e = \hbar = 1$, with m and e the electron rest mass and electric charge, respectively. In atomic units, energy is measured in hartree (Ha), where 1 Ha = 27.2 eV.

19.2 Formal Scattering Theory: The Problem

All too often the term *plane wave Born approximation* (PWBA) or, simply, Born approximation is encountered in theoretical treatments of the inelastic scattering of charged particles in matter. In fact, the PWBA represents the standard theoretical framework for studying the inelastic scattering of fast charged particles regardless of whether the target is in the condensed or gas phase. To understand why the PWBA is so popular we need to first understand why finding an exact solution to the formal scattering problem is impossible.

Inelastic scattering can be described as follows: a charged particle (considered structureless) of initial momentum $q_0 = \hbar k_0$ is scattered by a target atom to a state with momentum $q_n = \hbar k_n$ while the atom is excited from its ground state 0 to a state n. The energy transfer in the collision is $E_n = \hbar \omega_n$ (measured from state 0) and the transferred momentum $q = q_0 - q_n$.[*] Neglecting the effects of relativity and spin, it can be shown that the differential inelastic scattering cross section $d\sigma_{0n}/d\Omega$ is

$$\frac{d\sigma_{0n}}{d\Omega} \propto |\mathrm{M}_{0n}|^2 \tag{19.1}$$

where $d\Omega = 2\pi \sin\theta\, d\theta$, with θ being the scattering angle, which determines the direction of q_n relative to q_0, and M_{0n} the matrix element for the $0 \rightarrow n$ transition. The latter is given by

$$\mathrm{M}_{0n} \equiv \int \varphi_n^*(r)\Phi_n^*(r_1,\ldots,r_Z) \times V(r,r_1,\ldots,r_Z)\Psi^+(r,r_1,\ldots,r_Z)\,dr\,dr_1\ldots dr_Z$$

$$\equiv \left\langle \varphi_n \Phi_n \left| V \right| \Psi^+ \right\rangle \tag{19.2}$$

---

[*] We assume throughout that q is scalar in line with the approximation of amorphous (non-crystalline) materials, such as, for example, liquid water.

where $\varphi_n(r)$ and $\Phi_n(r_1,\ldots,r_Z)$ are the *final* state wavefunctions of the particle (at position r) and the target atom (with $r_1,\ldots,r_Z$ the coordinates of the Z atomic electrons), respectively, $\Psi^+(r,r_1,\ldots,r_Z)$ is the final state wavefunction of the *interacting* system (particle and target atom together) to be discussed below, V is the Coulomb potential for the particle-target atom interaction, and the asterisk denotes the complex conjugate of the wavefunction. We should note that owing to the fact that the electronic and nuclear motions are well separated (Born-Oppenheimer approximation), the atomic wavefunction is written as a product of an *electronic* wavefunction (i.e., the Φ in the above notation), which depends on the coordinates of the atomic electrons and a *nuclear* wavefunction, which depends on the coordinates of the nuclei. Given that we are here interested in the electronic excitations of the system, we consider only the electronic part of the atomic wavefunction. The Coulomb potential describing the interaction of the charged particle with the target electrons reads:

$$V = -\sum_{j=1}^{Z} \frac{ze^2}{|r-r_j|} \tag{19.3}$$

with ze being the particle charge and Ze the charge of the atomic nucleus. Then, the wavefunction Ψ^+ of Equation (19.2) can be found from the asymptotic boundary condition of the solution of the following Schrödinger equation:

$$(\tilde{H}_0 + \tilde{T} + V)\Psi = E\Psi \tag{19.4}$$

where $\tilde{H}_0$ is the Hamiltonian operator of an isolated atom, $\tilde{T}$ is the kinetic energy operator for the incident particle, V is the interaction Coulomb potential of Equation (19.3), and Ψ is the *total* (before and after scattering) wavefunction of the interacting system. The latter must obey the asymptotic boundary condition:

$$\Psi_{|r-r_j|\to\infty} \to \varphi_0\Phi_0 + \Psi^+ \tag{19.5}$$

where φ_0, Φ_0 are the *initial* state wavefunctions of the particle and atom, respectively. Thus, in formal scattering theory the inelastic scattering cross section depends, through the transition matrix element M_{0n}, upon the *final* state wavefunction of the *interacting* system (particle and target) Ψ^+, which must be found by solving Equation (19.4) under the asymptotic boundary condition of Equation (19.5). The problem is that an exact solution of Equation (19.4) is not possible because the interaction potential V depends upon the particle and target electron coordinates (r, r_j) appearing in the wavefunction Ψ, which is what we are trying to determine in the first place. Note that this conclusion is phase independent; i.e., it holds for both the gas and the condensed phase. Thus, since a formal evaluation of M_{0n} is not possible, one has to look for approximate solutions.

19.3 Born Approximation

The plane wave Born approximation (PWBA) provides a very useful approximate solution to the formal scattering problem by introducing two important simplifications to M_{0n}. First, by applying first-order perturbation theory, i.e., assuming that the interaction V is weak compared to $\tilde{H}_0$ and can be treated as a small perturbation, it replaces the *final* state wavefunction of the *interacting* system, Ψ^+, by the *initial* state wavefunction of the (noninteracting) system:

$$\Psi^+ \overset{\text{FBA}}{\to} \varphi_0 \Phi_0 \tag{19.6}$$

This assumption represents the *first Born approximation* (FBA). Then, the transition matrix element M_{0n} of Equation (19.2) becomes

$$M_{0n}^{\text{FBA}} = \langle \varphi_n \Phi_n | V | \varphi_0 \Phi_0 \rangle \tag{19.7}$$

Second, it assumes that the particle both before and after scattering can be described by plane waves:

$$\varphi_{0,n} = \exp(ik_{0,n}r) \tag{19.8}$$

The assumptions described by Equations (19.6) and (19.8) constitute the *plane wave Born approximation* (PWBA). To find the corresponding expression for M_{0n}^{PWBA} we insert Equation (19.8) into Equation (19.7) and, after using the relation (due to Bethe) $\int |r - r_j|^{-1} \exp(ikr)dr = (4\pi/k^2)\exp(ikr_j)$, where $k = q/\hbar$ is the transferred wavenumber, we obtain:

$$M_{0n}^{\text{PWBA}} = (ze^2) \frac{4\pi}{k^2} \langle \Phi_n | \sum_{j=1}^{Z} \exp(ikr_j) | \Phi_0 \rangle \tag{19.9}$$

By inspection of Equations (19.2) and (19.9) we can see that, whereas the transition matrix element M_{0n} of the exact solution depends upon the wavefunctions of the *interacting* system (particle and target), the M_{0n}^{PWBA} depends only upon the wavefunctions of the target (before and after the scattering). Thus, the PWBA reduces the calculation of inelastic scattering cross sections to a problem of determining the ground and excited states of the target— i.e, a typical *spectroscopy* problem. Note that the choice of plane waves in Equation (19.8) follows naturally from the asymptotic boundary condition Equation (19.5), since for large particle-atom separation the interaction potential V will be negligible and the incoming and scattered waves can then be described by plane waves. Therefore, quite often in the literature the result of Equation (19.9) is denoted as the first Born approximation, or simply the *Born approximation*, without reference to the assumption of plane waves.

19.3.1 Validity Range

Naturally, the PWBA would be reliable only for sufficiently *fast* charged particles, which would be minimally perturbed by the inelastic collision with the target, thus justifying both the weak interaction and plane wave assumptions. Although it is not clear how large the incident particle energy must be to validate the PWBA, it is generally assumed that the particle's velocity (v) should well exceed that of the target electrons. Given that the fastest electrons of an atom belong to the K-shell having a velocity equal approximately to $v_K \approx Z v_B$, with v_B being the Bohr velocity, the validity condition for the PWBA can be expressed as follows:

$$v \gg Z v_B \Rightarrow \begin{cases} T(\text{eV}) \gg 13.6\,Z^2 & \text{for electrons/positrons} \\ T_{ion}(\text{keV/amu}) \gg 25\,Z^2 & \text{for ions} \end{cases} \tag{19.10}$$

For water ($Z = 10$) the above conditions result in a lower limit of about 1.3 keV for electrons and 2.5 MeV/amu for ions. In practice, it has been found that for calculating the *total* inelastic scattering cross sections (or, equivalent, the inverse inelastic mean free path) the PWBA is valid down to much lower energies than the conditions of Equation (19.10) seem to imply. This is not unexpected, since Equation (19.10) is based on the velocity of the K-shell electrons, which are the most energetic, while contributing little to inelastic scattering compared to the valence electrons. Thus, PWBA calculations are generally considered quite reliable for electrons down to ~300 eV and for protons (or other light ions) down to ~500 keV/amu. Semiquantitative results, though, can often be obtained down to much lower energies (e.g., 50–100 eV for electrons).

Although the PWBA has simplified matters considerably by transforming the formal scattering problem whereby one needs to calculate the wavefunctions of the *interacting* system to a pure spectroscopy problem of determining the electronic energy structure of the target, the actual calculations are far from trivial. This is because the calculation of good quality excited state wavefunctions (Φ_n) of a many-electron system (being an atom, a molecule, or a solid) needed to obtain M_{0n}^{PWBA} is a formidable task. Alternative approximations to the formal scattering solution, which are also based on the FBA, such as the Coulomb wave Born approximations (CWBAs) or the distorted wave Born approximation (DWBA) that make use of more realistic descriptions of the particle wavefunctions ($\varphi_{0,\,n}$) and are therefore more accurate than the PWBA, are too complicated to be useful for practical calculations; e.g., the transition matrix element depends upon both the particle and target wavefunctions.

19.3.2 Dynamic Structure Factor

An important consequence of the form of M_{0n}^{PWBA} as expressed in Equation (19.9) is that the differential inelastic scattering cross section in PWBA can be factorized into a *particle-dependent* (kinematic) factor and a *material-dependent* (dynamic) factor. The complete expression for the inelastic scattering cross section differential in the transferred wavenumber can be written as follows:[*]

$$d\sigma_n^{PWBA}/dk = \underbrace{(8\pi e^4/\hbar^2)}_{\text{constant}} \underbrace{(z^2/\upsilon^2)}_{\text{particle}} \underbrace{|F_n(k)|^2}_{\text{target}} (1/k^3) \tag{19.11}$$

where $|F_n(k)|^2$ is the absolute square of the following transition matrix element:

$$F_n(k) \equiv \langle \Phi_n | \sum_{j=1}^{Z} \exp(ikr_j) | \Phi_0 \rangle \tag{19.12}$$

Comparing Equations (19.9) and (19.12), we obtain:

$$M_n^{PWBA} = (ze^2) \frac{4\pi}{k^2} F_n(k) \tag{19.13}$$

Apparently, $|F_n(k)|^2$ is the only nontrivial factor in Equation (19.11). The physical meaning of $|F_n(k)|^2$ is that it represents the conditional probability that the target is excited to state n upon receiving momentum transfer $q = \hbar k$. Thus, it reflects all the dynamic properties of matter, and therefore it is called the *dynamic structure factor* (also known as the inelastic form factor) of the material.[†] The definition of $|F_n(k)|^2$ implies that, contrary to a two-body (i.e., binary) collision where there is a one-to-one correspondence between momentum transfer and energy transfer, this correspondence does not any longer hold in a many-body system, both classically and quantum mechanically. Specifically, in quantum mechanics the probabilistic connection between energy and momentum transfer is due to the fact that the target electron does not have a well-defined momentum before scattering but a continuous probability distribution dictated by its wavefunction. Also, classically, a momentum transfer to a *bound* electron still allows the rest of the atom to recoil in an unspecified way, thus rendering the excitation energy of (or equivalently, the energy transfer to) the system a priori indeterminate. However, it is only in quantum mechanics that a precise way to determine $|F_n(k)|^2$ is available through Equation (19.12) that entails the ground and excited state wavefunctions of the target. The fact that $|F_n(k)|^2$ is independent of the charge, mass, or velocity of the particle has an important practical implication. If one can measure $d\sigma_n/dk$ at a *single particle energy* (sufficiently large for the PWBA to be valid), then Equation (19.11) can be used to infer $|F_n(k)|^2$, which can be used in Equation (19.11) to calculate $d\sigma_n/dk$ for *any other particle type and energy*.

---

[*] The shortened notation n instead of $0n$ will hereafter be used.
[†] Sometimes the dynamic structure factor is identified with $F_n(k)$ and not $|F_n(k)|^2$.

In principle, the dynamic structure factor is all that is needed for inelastic scattering calculations within PWBA in both the gas and condensed phases. However, as with M_n^{PWBA}, the calculation of the dynamic structure factor on the basis of its definition, Equation (19.12), meets with the formidable task of determining the excited state wavefunctions of the system. Therefore, in practice, it has been found most convenient to replace the dynamic structure factor by two other materials functions: the oscillator strength and the dielectric response function, which, although essentially equivalent to it, are easier to interpret, and most importantly, they have a more straightforward connection to various *measurable* properties of materials.

19.3.3 Oscillator Strength

For isolated atoms or molecules (i.e., for the gas phase) it is customary to express the dynamic structure factor by a slightly different quantity, namely the atomic (or molecular) *generalized oscillator strength* (GOS) first introduced by Bethe and denoted by $f_n(Q)$. The two quantities are related through

$$f_n(Q) = \frac{E_n}{Q} |F_n(k)|^2 \qquad (19.14)$$

where

$$Q = q^2/2m \qquad (19.15)$$

with m being the electron rest mass and Q the so-called recoil energy, i.e., the kinetic energy of a free and stationary electron after receiving a momentum q. The recoil energy can also be written as $Q = Ry\,(k\,a_B)^2$, where Ry stands for the Rydberg constant $(1\,Ry = \frac{me^4}{2\hbar^2} = 13.6\,eV)$ and a_B is the Bohr radius $(a_B = \frac{\hbar^2}{me^2} = 0.529\,\text{Å})$.

The GOS is a direct extension to arbitrary momentum transfer ($q \neq 0$) of the more familiar optical oscillator strength (OOS) used in optical spectroscopy. The two quantities are related through

$$f_n = \lim_{Q \to 0} f_n(Q) \qquad (19.16)$$

where $f_n \equiv f_n(Q=0)$ is the OOS. To obtain a more informative form of the OOS we must examine $f_n(Q)$ in the limit $k \to 0$, the so-called optical[*] or long-wavelength[†] limit that pertains to the nearly forward scattering (i.e., at

---

[*] The name originates from the fact that optical photons transfer negligible momentum to the system.
[†] The magnitude of the momentum transfer q is indicative of the length scale of the perturbation induced in the material by the charged particle through $q = \hbar k = h/\lambda$, where λ is the wavelength of the induced charge density oscillation.

vanishingly small angles)* of charged particles. For $Q \ll 1$ (i.e., small values of k) we can expand the exponential term in Equation (19.12) according to $\exp(ikr_j) \approx 1 + ikr_j$. Then, using Equations (19.14) and (19.15) we obtain

$$\lim_{Q \to 0} f_n(Q) = \frac{E_n}{Q} \left[|\langle \Phi_n | 1 | \Phi_0 \rangle|^2 + k^2 |\langle \Phi_n | \sum_{j=1}^{Z} r_j | \Phi_0 \rangle|^2 + \cdots \right] \approx (E_n/Ry) |D_n|^2 \quad (19.17)$$

where $|\langle \Phi_n | 1 | \Phi_0 \rangle|^2 = 0$ due to the orthogonality of the wavefunctions and

$$D_n \equiv \langle \Phi_n | \sum_{j=1}^{Z} r_j | \Phi_0 \rangle / a_B \quad (19.18)$$

defines the so-called *dipole matrix element*. The dipole matrix element D_n determines the photoabsorption cross section σ_n^{ph} via

$$\sigma_n^{ph} = 4\pi^2 \alpha \, a_B^2 \, |D_n|^2 \quad (19.19)$$

with $\alpha = \frac{e^2}{\hbar c} \approx \frac{1}{137}$ being the fine-structure constant (c is the speed of light). Then, with the aid of Equations (19.16) and (19.17), the OOS becomes

$$f_n = \frac{E_n}{Ry} |D_n|^2 \quad (19.20)$$

By inserting Equation (19.20) into Equation (19.19) we obtain the important result:

$$\sigma_n^{ph} = 4\pi^2 \alpha \, a_B^2 Ry \, (f_n/E_n) \quad (19.21)$$

Thus, the cross section for the absorption of photons, which can be measured by standard spectroscopic techniques, is proportional to the OOS.

An important property of the OOS is that its sum over all possible excitation states n gives the total number of electrons in the target atom or molecule:[†]

$$\sum_n f_n = Z \quad (19.22)$$

This is the so-called *f-sum rule* (also known as the Thomas-Reich-Kuhn sum rule). Bethe showed that for any *fixed* value of Q, the *f*-sum rule also holds for

---

* There is a one-to-one correspondence between momentum transfer and scattering angle through Equation (19.43).

† For Equation (19.22) to hold, f_n must be expressed per target atom (or molecule). If, on the other hand, f_n is expressed per target electron, then the value of the *f*-sum rule adds to unity instead of Z (see also later).

the GOS; that is,

$$\sum_n f_n(Q) = Z \qquad (19.23)$$

The discussion so far has implicitly assumed that the excitation state n is discrete. For excitation to states in the continuum (i.e., ionizations in the gas phase or excitations to the conduction band of condensed media) it is more appropriate to represent the excitation energy E_n (or energy transfer) by a continuous variable symbolized by W. We can then define the differential distribution of OOS (or GOS) per unit range of W (per atom or molecule). The relation between the photoabsorption cross section and the OOS, Equation (19.21), should then be recast as

$$\sigma_{ph} = 4\pi^2 \alpha \, a_B^2 \mathrm{Ry} \, df(W)/dW \qquad (19.24)$$

where $df(W)/dW \equiv df(W, Q = 0)/dW$. The f-sum rule (for any Q, including $Q = 0$) then reads:

$$\int_0^\infty \frac{f(W, Q)}{dW} dW = Z \qquad (19.25)$$

An important aspect of the f-sum rule is that if the upper limit of integration in Equation (19.25) is replaced by W', then the f-sum rule counts the *effective* number of target electrons Z_{eff} that participate in inelastic collisions with energy transfer up to W'. This is sometimes called *partial f*-sum rule and expressed as

$$\int_0^{W'} \frac{f(W, Q)}{dW} dW = Z_{eff}(W') \qquad (19.26)$$

The double-differential (in energy and momentum transfer) inelastic cross section can then be expressed as follows:

$$d^2\sigma_{PWBA}/dWdQ = \underbrace{(2\pi e^4/m)}_{\text{constant}} \underbrace{(z^2/v^2)}_{\text{particle}} \underbrace{df(W, Q)/dW}_{\text{target (GOS)}} (1/WQ) \qquad (19.27)$$

Thus, knowledge of GOS over the entire range of energy and momentum transfer is all that is needed for calculating the inelastic scattering cross section (in PWBA) as a function of particle velocity. To highlight the importance of this fact, the surface represented by the GOS over the W-Q plane is given the special name *Bethe surface* of the atom (or molecule). For $Q \approx 0$ the maxima of the Bethe surface correspond to the various discrete excitation transitions and binding energies of individual shells of the atom or molecule. With increasing Q the maxima shift to higher energies with a concomitant decrease in their amplitude. Eventually, for values of Q much larger than the binding energies of the target electrons, the excitation peaks merge along the line $W \approx Q$; i.e., the most probable energy transfer approaches the value of the recoil energy Q. The region of the Bethe surface whereby the GOS peaks along the line $W \approx Q$ is called the *Bethe ridge*.

For a two-body collision, where the struck electron is free and stationary, the GOS takes the form of a delta-function:

$$\frac{df(W,Q)}{dW} = Z\delta(W - Q) \tag{19.28}$$

which implies a quadratic relation between energy and momentum transfer and a sharply peaked Bethe ridge of zero width. However, in real systems, the Bethe ridge has a nonzero width (i.e., it is broadened) due to binding effects. Thus, it is often said that many-body effects cause a smearing on the δ-like profile of the GOS over the energy momentum plane. Close analytic forms of GOS are available only for the hydrogen atom and the homogeneous electron gas (to be discussed in Section 19.5). The GOS for the hydrogen atom (see Egerton 1996) is shown in Figure 19.1 both as a three-dimensional (3D) surface (a) and as two-dimensional (2D) contour plot (b). The emergence of a Bethe ridge of finite width with increasing W and q is evident. The numerical calculation of the complete Bethe surface from first principles for other atomic systems is far from trivial (e.g., see Segui et al. 2002). We should note in passing that a two-body collision is formally equivalent to an *elastic* scattering since, in this case, the target (i.e., the scattering electron) lacks an internal energy structure. In this case the inelastic scattering is exactly described by the Rutherford formula (both classically and quantum mechanically), as one can trivially verify by inserting Equation (19.28) into Equation (19.27) and integrate over Q.

19.3.4 Dielectric Response Function

For condensed media it is convenient to use yet another material-dependent function by replacing the dynamic structure factor by the *dielectric response function* (or simply dielectric function), which is a generalization of the dielectric constant ε_0 of matter. In simple words, the dielectric constant ε_0 is the factor by which, due to the polarization of the medium, an applied electric field is decreased to yield the net internal field in the medium. Its generalization to a time- and space-varying perturbation yields a complex-valued dielectric constant, which is now a function of frequency $\omega = W/\hbar$ and wavenumber $k = q/\hbar$:

$$\varepsilon(W,q) = \varepsilon_1(W,q) + i\varepsilon_2(W,q) \tag{19.29}$$

with $\varepsilon_1(W,q) \equiv \mathrm{Re}\,\varepsilon(W,q)$ and $\varepsilon_2(W,q) \equiv \mathrm{Im}\,\varepsilon(W,q)$ being, respectively, the real and imaginary parts. Similar to the GOS, the dielectric function provides

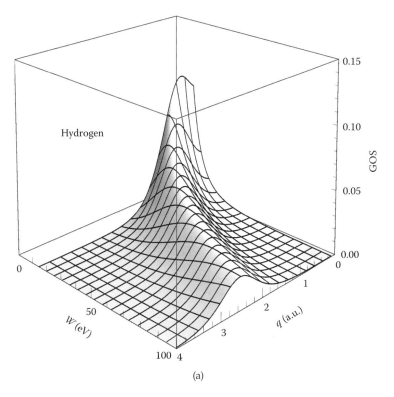

(a)

FIGURE 19.1

The generalized oscillator strength (GOS) of the hydrogen atom as a function of energy transfer W and momentum transfer q depicted as a 3D surface, the so-called Bethe surface (a) and a 2D contour plot (b).

an important bridge between many *measurable* properties of the materials and their electronic response to external charged particles.

To see now how the dielectric function enters into the present formalism we should note two things. First, the imaginary part $\varepsilon_2(W,q)$ corresponds, in the limit of a vanishing screening (i.e., in the gas phase), to the single-electron excitations of the system. Thus, $\varepsilon_2(W,q)$ is directly related to the GOS through

$$\varepsilon_2(W,q) = 2\pi^2 e^4 \alpha_B \, NZ \frac{1}{W} \frac{df(W,Q)}{dW} \tag{19.30}$$

where N is the atomic density of the material (this makes the product NZ equal the electronic density), and $df(W,Q)/dW$ is the GOS *per electron* (also called one-electron GOS), which is conceptually more appropriate when discussing condensed phase properties than the atomic GOS (i.e., the GOS *per atom*). Through Equations (19.24) and (19.30), $\varepsilon_2(W,q=0)$ can also be obtained from the photoabsorption cross section. Contrary to the atomic

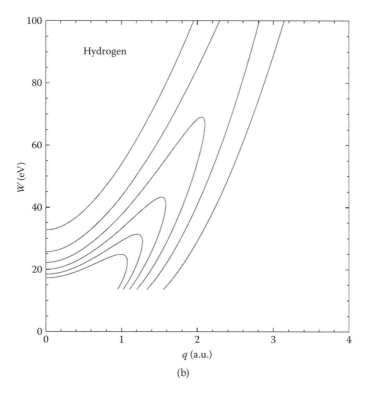

FIGURE 19.1
(Continued)

GOS, the *f*-sum rule of the one-electron GOS adds to unity and not to Z. So, using Equations (19.25) and (19.30) we can write the following *f*-sum rule for the imaginary part of the dielectric function:

$$\frac{2}{\pi E_{pl}^2} \int_0^\infty W \varepsilon_2(W,q)dW = 1 \qquad (19.31)$$

where

$$E_{pl} = \sqrt{4\pi \hbar^2 e^2 N\, Z/m} \qquad (19.32)$$

defines the oscillation energy of a "gas" of free electrons (also called plasmon oscillation; to be discussed later) with electronic density equal to NZ. In analogy to the partial *f*-sum rule for GOS, Equation (19.26), we may also define the corresponding partial *f*-sum rule for $\varepsilon_2(W,q)$ by integrating Equation (19.31) up to a finite value W' instead of infinity.

    Second, we should note that, contrary to the gas phase where $|\varepsilon(W,q)| = 1$, in the condensed phase $|\varepsilon(W,q)| \neq 1$; that is, the Coulomb interaction between the charged particle and the target electrons is screened (rather than bare), so the Coulomb potential of Equation (19.3) is modified as follows:

$$V_{scr} = \frac{V}{|\varepsilon(W,q)|} \tag{19.33}$$

This screened Coulomb potential is part of the transition matrix element of Equation (19.2), and therefore $|\varepsilon(W,q)|$ will carry through as $|\varepsilon(W,q)|^2$ in subsequent calculations (which involve the *square* of the transition matrix element). Thus, screening effects in the condensed phase will modify the spectrum of the single-electron excitations of the gas phase by a factor of $|\varepsilon(W,q)|^2$ in the denominator via the replacement:

$$\underbrace{\varepsilon_2(W,q)}_{\text{gas phase}} \rightarrow \underbrace{\varepsilon_2(W,q)/|\varepsilon(W,q)|^2}_{\text{condensed phase}} \tag{19.34}$$

where the right-hand side of Equation (19.34) is the so-called *energy loss function* (ELF)[*]:

$$\underbrace{\mathrm{Im}\left[\frac{-1}{\varepsilon(W,q)}\right]}_{\text{ELF}} = \frac{\varepsilon_2(W,q)}{|\varepsilon(W,q)|^2} \tag{19.35}$$

Based on Equations (19.30) and (19.35) we can extend the concept of GOS to the condensed phase and write

$$\mathrm{Im}\left[\frac{-1}{\varepsilon(W,q)}\right] = \frac{\pi E_{pl}^2}{2} \frac{1}{W} \frac{df(W,Q)}{dW} \tag{19.36}$$

where E_{pl} is given by Equation (19.32). The one-electron GOS, Equation (19.36), now includes the effects of screening, contrary to Equation (19.30). In analogy to Equation (19.31) we can write an *f*-sum rule for the ELF as follows:

$$\frac{2}{\pi E_{pl}^2} \int_0^\infty W \, \mathrm{Im}\left[\frac{-1}{\varepsilon(W,q)}\right] dW = 1 \tag{19.37}$$

A partial *f*-sum rule for the ELF may be defined by substituting the infinity by W' as the upper integration limit in Equation (19.37).

---

[*] Sometimes called dielectric loss function or simply loss function

19.3.5 Kramers-Krönig Relations

In experiments one commonly measures either the real or the imaginary parts of $\varepsilon(W,q)$ or of $1/\varepsilon(W,q)$. The Kramers-Krönig (KK) relations show that the real and imaginary parts are not independent, so one can determine one from the other. With respect to the dielectric function we have

$$\varepsilon_1(W,q) = 1 + \frac{1}{\pi} P \int_{-\infty}^{\infty} \varepsilon_2(W',q) \frac{dW'}{W' - W} \tag{19.38a}$$

$$\varepsilon_2(W,q) = -\frac{1}{\pi} P \int_{-\infty}^{\infty} [\varepsilon_1(W',q) - 1] \frac{dW'}{W' - W} \tag{19.38b}$$

where the symbol P stands for the Cauchy principal value avoiding the pole at $W = W'$. Consequently, if $\varepsilon_1(W,q)$ or $\varepsilon_2(W,q)$ is known, then the other can be determined through the KK relations. In analogy to Equations (19.38) we also have

$$\mathrm{Re}\left[\frac{1}{\varepsilon(W,q)}\right] = 1 + \frac{2}{\pi} P \int_0^{\infty} \mathrm{Im}\left[\frac{1}{\varepsilon(W',q)}\right] \frac{W'dW'}{W^2 - W'^2} \tag{19.39a}$$

$$\mathrm{Im}\left[\frac{1}{\varepsilon(W,q)}\right] = \frac{2}{\pi} P \int_0^{\infty} \left\{1 - \mathrm{Re}\left[\frac{1}{\varepsilon(W',q)}\right]\right\} \frac{W'dW'}{W^2 - W'^2} \tag{19.39b}$$

One can then use $1/\varepsilon(W,q)$ to obtain $\varepsilon(W,q)$ from the relations:

$$\varepsilon_1(W,q) = \frac{\mathrm{Re}[1/\varepsilon(W,q)]}{\{\mathrm{Re}[1/\varepsilon(W,q)]\}^2 + \{\mathrm{Im}[1/\varepsilon(W,q)]\}^2} \tag{19.40a}$$

$$\varepsilon_2(W,q) = \frac{\mathrm{Im}[1/\varepsilon(W,q)]}{\{\mathrm{Re}[1/\varepsilon(W,q)]\}^2 + \{\mathrm{Im}[1/\varepsilon(W,q)]\}^2} \tag{19.40b}$$

19.3.6 Dielectric Formulation

Using Equation (19.36) it is now straightforward to recast the expression for the differential inelastic scattering cross section, Equation (19.27), in a form appropriate to the condensed phase that incorporates the dielectric function via the ELF. However, for the condensed phase there is a change in nomenclature. The standard (microscopic) cross section in units of area per target atom or molecule, commonly used for the gas phase, is now replaced by what is called a "macroscopic" cross section or, equivalent, an *inverse inelastic mean free path*, which has units of reciprocal length. The reason for this change is that in the condensed phase the charged particle will interact simultaneously with a large number of electrons belonging to different atoms or molecules; in fact, this is exactly what the dielectric function is meant to describe. Therefore, contrary to the gas phase, it is misleading to associate an inelastic cross section with an individual atom or molecule, but it is rather more appropriate to use a macroscopic inelastic cross section associated with the probability of interaction per unit track length in the medium. Thus, we write for Λ, the inverse inelastic mean free path in units of length:

$$\Lambda = N\sigma \tag{19.41}$$

where σ is the microscopic cross section in units of area/atom, and N is the atomic density in units of atoms/volume. Then, the double-differential inverse inelastic mean free path reads:

$$d^2\Lambda_{PWBA}/dWdq = \underbrace{(2/\pi a_B m)}_{\text{constant}}\underbrace{(z^2/v^2)}_{\text{particle}}\underbrace{\text{Im}[-1/\varepsilon(W,q)]}_{\text{target (ELF)}}(1/q) \tag{19.42}$$

For incident electrons with kinetic energy T, it is sometimes useful to recast Equation (19.42) in terms of the electron scattering angle θ using the following relation between q and θ:

$$q = \sqrt{2m}\,[2T - W - 2\sqrt{T(T-W)}\cos\theta]^{1/2} \tag{19.43}$$

Then, the double-differential inverse inelastic mean free path for incident electrons ($z^2 = 1$) becomes

$$d^2\Lambda_{PWBA}/dWd\theta = (2/a_B T)\,\text{Im}[-1/\varepsilon(W,q)]\,g(W,\theta) \tag{19.44}$$

where

$$g(W,\theta) = \frac{\sqrt{T(T-W)}\sin\theta}{2T - W - 2\sqrt{T(T-W)}\cos\theta} \tag{19.45}$$

In PWBA all relevant magnitudes follow from the double-differential inelastic cross section by suitable integrations over the W and q variables. For example, the single differential in the energy transfer inverse inelastic mean free path is found by integrating Equation (19.42) over q:

$$\Lambda_W \equiv \frac{d\Lambda}{dW} = \int_{q_{min}}^{q_{max}}\frac{d^2\Lambda}{dWdq}dq = \frac{z^2}{\pi a_B T}\int_{q_{min}}^{q_{max}}\text{Im}\left[\frac{-1}{\varepsilon(W,q)}\right]\frac{dq}{q} \tag{19.46}$$

where the integration limits for incident electrons (and positrons) are

$$q_{max} = \sqrt{2m}(\sqrt{T} + \sqrt{T-W}) \quad \text{and} \quad q_{min} = \sqrt{2m}(\sqrt{T} - \sqrt{T-W}) \tag{19.47}$$

For incident ions the expressions of Equation (19.47) hold with T replaced by T_{ion} and m by m_{ion} where $T - T_{ion} \times m/m_{ion}$.

The total inverse inelastic mean free path is then found by integrating Equation (19.46) further over W:

$$\Lambda_{\text{tot}} \equiv \int_{W_{\text{min}}}^{W_{\text{max}}} \frac{d\Lambda}{dW} dW = \frac{z^2}{\pi a_B T} \int_{W_{\text{min}}}^{W_{\text{max}}} dW \int_{q_{\text{min}}}^{q_{\text{max}}} \text{Im}\left[\frac{-1}{\varepsilon(W,q)}\right] \frac{dq}{q} \qquad (19.48)$$

where the upper limits of integration are commonly evaluated assuming binary collisions, i.e., $W_{\text{max}} \approx T/2$ for electrons[*], $W_{\text{max}} \approx T$ for positrons, and $W_{\text{max}} \approx 4T$ for ions. For the low limit of integration we have $W_{\text{min}} = E_{\text{gap}}$ for insulators and semiconductors and $W_{\text{min}} = E_F$ for conductors, with E_{gap} and E_F being the band-gap and Fermi energies, respectively. Finally, the electronic (or collision) stopping power can be obtained from

$$\Lambda_{\text{st}} \equiv \int_{W_{\text{min}}}^{W_{\text{max}}} W \frac{d\Lambda}{dW} dW = \frac{z^2}{\pi a_B T} \int_{W_{\text{min}}}^{W_{\text{max}}} W dW \int_{q_{\text{min}}}^{q_{\text{max}}} \frac{1}{q} \text{Im}\left[\frac{-1}{\varepsilon(W,q)}\right] dq \qquad (19.49)$$

It is clear from the above expressions that knowledge of the ELF of the material as a function of W and q completely determines (within the PWBA) the inelastic scattering and (electronic) energy loss of charged particles in condensed media. Therefore, in analogy to the gas phase, the three-dimensional plot of ELF over the $W - q$ plane is called the *Bethe surface* of the material. We may note in passing that the origin of the name energy loss function follows from the fact that, by virtue of Equation (19.42) or (19.44), the ELF is proportional to the *energy loss spectrum* (at a given particle velocity and scattering angle) in a transmission experiment. Note, however, that the ELF is solely a material property independent of the projectile, whereas the energy loss spectrum does depend upon the (velocity and type of) projectile.

Similar to the GOS, the dielectric function (or ELF) is formally obtained from the ground and all excited state wavefunctions of the system through the dynamic structure function. A closed analytical expression for the ELF exists only for the idealized case of the homogeneous electron gas (to be discussed in Section 19.5). The calculation of the ELF from first principles for realistic materials is difficult (Zaider 1991). At present, *ab initio* time-dependent density functional theory (TDDFT) calculations represent the most sophisticated method for obtaining $\varepsilon(W, q)$. TDDFT is basically a generalization of DFT (a theory for ground state properties) to excited states. The problem with TDDFT in the present context is that, due to computational difficulties, calculations are limited to small values of the parameters (W, q) and, as a result, can be mostly useful for very low-energy incident particles, e.g., electrons with energies close to the band-gap (or the Fermi energy).

---

[*] The division by 2 arises from the indistinguishability of the incident and target electrons and the convention to assume that the more energetic electron after the collision is the primary one.

From the experimental side the ELF can be determined by irradiating the material with a charged particle beam (most often an electron beam is used) and analyzing, via Equation (19.44), the outgoing particle spectra at different scattering angles corresponding to different momentum transfer. With such experiments, however, it is not possible to probe the entire Bethe surface due to multiple scattering effects at large q, since large-angle scattering will likely cause the incident charged particles to experience more than one interaction within the material.

Therefore, for practical calculations of the inelastic scattering of relatively high-energy charged particles whereby the ELF must be known over a broad range of energy and momentum transfers, one is left with the following two choices: (1) first principles model calculations using the homogeneous electron gas (HEG) theory, or (2) semiempirical calculations using optical data models (ODMs). However, before proceeding to the discussion of HEG theory and ODMs, it is important that we discuss first yet another approximation (the Bethe approximation), which reveals that at sufficiently high particle energies knowledge of ELF at $q \approx 0$ (optical limit) will suffice.

19.4 Bethe Approximation

The PWBA is a high-energy approximation valid for large T in the above notation, where T is proportional to the square of particle velocity. The Bethe theory provides an asymptotic expansion of the PWBA in powers of T^{-1}. The advantage of such an expansion is that it is then straightforward to recognize terms with increasing contribution at high T (in PWBA). Importantly, it turns out that the expansion coefficients are uniquely determined from the ELF. The Bethe asymptote, which holds for any momentum transfer integrated cross section in PWBA (e.g., for Λ_W, Λ_{tot}, and Λ_{st}) has the general form (Inokuti 1971)

$$\Lambda = AT^{-1}\ln(T/\mathrm{Ry}) + BT^{-1} + CT^{-2} + O(T^{-3}) \tag{19.50}$$

where $O(T^{-3})$ denotes terms of order T^{-3} and higher. Obviously, the magnitude of the Bethe asymptote at large particle velocities ($T \gg \mathrm{Ry}$) is mainly determined from the value of the coefficient A of the logarithmic term, while the details of its variation with T will depend upon the coefficient B. At lower particle velocities the coefficient C of the term of order T^{-2} will also become important. Given that the PWBA is a high-energy approximation, the usefulness of the Bethe asymptote is expected to be practically exhausted by the first three terms; that is, the terms $O(T^{-3})$ will become sizable at such low T where the PWBA (and Bethe theory) would be invalid. Hence, the extension of the Bethe asymptote to terms of order T^{-3} and higher is practically meaningless.

An important outcome of the Bethe theory is that the coefficient A of the leading logarithmic term that dominates at high T is determined *solely* from the optical limit $(q \to 0)$ of ELF in the following way:

$$A = \begin{cases} A_W = \text{const} \times \text{Im}\left[-1/\varepsilon(W, q = 0)\right] \\[2mm] A_{\text{tot}} = \text{const} \times \int \text{Im}\left[-1/\varepsilon(W, q = 0)\right] dW \\[2mm] A_{\text{st}} = \text{const} \times 2 \int W \, \text{Im}\left[-1/\varepsilon(W, q = 0)\right] dW \end{cases} \tag{19.51}$$

where $\text{const} = z^2/2\pi a_{\text{B}}$.

This finding is of practical importance because, as it will be discussed in Section 19.6.1, the optical limit of the ELF can be measured experimentally. The value of A_{st} that enters the stopping power can be further simplified by using the *f*-sum rule, Equation (19.37), to obtain

$$A_{\text{st}} = \frac{z^2 E_{pl}^2}{2 a_{\text{B}}} \tag{19.52}$$

where E_{pl} is given by Equation (19.32).

On the other hand, the coefficients B_W and B_{tot} are much more complicated and depend upon the ELF at $q \neq 0$. In contrast, via the action of the *f*-sum rule of Equation (19.37), the coefficient B_{st} of stopping power can also be shown to depend upon the optical limit of ELF as follows:

$$B_{\text{st}} = A_{\text{st}} \ln(2 \, \text{Ry}/I) \quad \text{(electrons/positrons)}$$
$$B_{\text{st}} = A_{\text{st}} \ln(4 \, \text{Ry}/I) \quad \text{(ions)} \tag{19.53}$$

where I denotes the so-called mean excitation energy of the material (a magnitude that pertains to stopping power theory). For condensed media, I is defined in terms of the ELF as follows:

$$\ln(I) \equiv \frac{\displaystyle\int_0^\infty W \ln(W) \, \text{Im}[-1/\varepsilon(W, q = 0)] dW}{\displaystyle\int_0^\infty W \, \text{Im}[-1/\varepsilon(W, q = 0)] dW} \tag{19.54}$$

Evidently, $\ln(I) \equiv \langle \ln(W) \rangle$, where the average is taken with respect to the OOS, since $df(W, q = 0)/dW \propto W \, \text{Im}[-1/\varepsilon(W, q = 0)]$.

Finally, the expressions for the coefficients C_W, C_{tot}, and C_{st} of the T^{-2} term are much more complicated and depend upon features of the ELF at $q \neq 0$. These coefficients are related to the so-called inner shell effects, which become important at relatively low incident particle velocities or, more specifically, at incident particle velocities comparable to those of the bound electrons. From Equation (19.50) Bethe's asymptotic expression of the PWBA stopping power reads:

$$\Lambda_{st}^{PWBA} \equiv -(dT/dx)_{PWBA} = \underbrace{A_{st}T^{-1}\ln(T/Ry) + B_{st}T^{-1} +}_{\text{Bethe}} \underbrace{C_{st}T^{-2}}_{\text{shell-correction}} + O_{st}(T^{-3})$$

(19.55)

Using now Equations (19.52) and (19.53) in Equation (19.55), we obtain the standard (uncorrected or asymptotic) expression of the Bethe stopping power formula:

$$\Lambda_{st}^{Bethe} = A_{st}T^{-1}\ln(T/Ry) + B_{st}T^{-1} = \begin{cases} \dfrac{z^2 E_{pl}^2}{2a_B T}\ln\left(\dfrac{2T}{I}\right) & \text{(electrons/positrons)} \\[3mm] \dfrac{z^2 E_{pl}^2}{2a_B T}\ln\left(\dfrac{4T}{I}\right) & \text{(ions)} \end{cases}$$

(19.56)

The coefficient C_{st} represents the so-called shell correction term to the Bethe stopping power formula, which becomes important at low particle velocities and, as mentioned above, depends upon the ELF at $q \neq 0$ (Fano 1963).

It is important to emphasize that the Bethe stopping power formula, Equation (19.56), depends solely upon the ELF at $q = 0$ via the I value defined by Equation (19.54). Thus, to order T^{-1}, differences between the gas and condensed phase (normalized to the same mass density), enter the stopping power through their different I values. In analogy to the *high-energy density effect* in the stopping power of condensed media to relativistic particles, which is related to the transverse component of the interaction, the above effect, relevant to nonrelativistic particles, is often called the *zero-energy* or *longitudinal density effect*.

In conclusion, the standard Bethe stopping power formula, Equation (19.56), is based not only upon the PWBA but also on an additional high-energy approximation that assumes that nonoptical terms (of order T^{-2} and higher) that depend upon the ELF at $q \neq 0$ are negligible; thus, it is often denoted as the *optical approximation* or, simply, the Bethe approximation.

19.5 Electron Gas Theory

The homogeneous electron gas (HEG) theory, or so-called jellium model, provides a very useful theoretical tool for studying electronic excitations in the condensed phase. The main assumption of the HEG theory is that the material is composed of free (i.e., unbound) electrons, the so-called electron gas that moves in a structureless, homogeneous, and isotropic background of positively charged ion cores that provide charge neutrality. In what follows we will use the term *electron* to refer to the *screening* electrons of the HEG and the term *ion* to refer to the lattice ion cores. The HEG theory is considered a first principles theory since it contains only a single free parameter, namely, the electronic density of the material, $N_e \equiv N \times Z$. It is customary to use, instead of N_e, an alternative dimensionless parameter, the so-called one-electron radius or electron gas parameter, defined as follows. The inverse of N_e by definition, gives the volume occupied by one electron, i.e., $V_e = 1/N_e$. The latter can be mathematically expressed through a sphere of radius R_e according to the relation $V_e = (4/3)\pi R_e^3$ or, after solving for R_e:

$$R_e = \left(\frac{3V_e}{4\pi}\right)^{1/3} = \left(\frac{3}{4\pi N_e}\right)^{1/3} \tag{19.57}$$

The dimensionless electron gas parameter, r_s, is then defined by the relation

$$r_s = R_e/a_B \tag{19.58}$$

Typical values of r_s for metals range between 2 and 6, whereas for organic materials they lie between 1.5 and 2. The value of r_s is the only free parameter in HEG theory.

EXAMPLE 19.1

Calculate the electron gas parameter for liquid water assuming a mass density of $\rho = 1\,\mathrm{g\,cm^{-3}}$.

SOLUTION 19.1

The electronic density of liquid water is (with A the atomic weight and N_{Av} the Avogadro constant)

$$N_e = \frac{\rho N_{Av} Z}{A} = \frac{1\,\mathrm{g\,cm^{-3}} \times 6.022 \times 10^{23}\,\mathrm{mol^{-1}} \times 10\,\mathrm{e^-}}{18\,\mathrm{g/mol}} = 3.34 \times 10^{23}\,\mathrm{e^-/cm^3}$$

Then, from Equation (19.57), we obtain

$$R_e = \left(\frac{3}{4 \times 3.14 \times 3.34 \times 10^{23} \times 10^{-24}\,\overset{o}{\mathrm{A}}^{-3}}\right)^{1/3} = 0.894\,\overset{o}{\mathrm{A}}$$

Substituting to Equation (19.58) we obtain

$$r_s = \frac{0.894}{0.529} = 1.69$$

19.5.1 Plasmons

The HEG theory is expected to be most suitable for the conduction electrons of metals or, more generally, for the weakly bound valence electrons of free-electron-like materials. The latter are defined as those materials where their ELF at $q \approx 0$ exhibits a well-defined peak at about the nominal plasmon energy E_{pl} given by Equation (19.32). This is indeed the case with many biological materials (liquid water, DNA, etc.), despite the fact that they are also wide band-gap insulators. This apparent contradiction can be explained by the relatively large difference between the plasmon energy and the band-gap energy $(E_{pl} \gg E_{gap})$ in these materials, since the larger the plasmon energy, the less sensitive the response of the system to details of its band-structure.

The HEG can exhibit two types of excitations, namely single-electron excitations (i.e., e-h pairs) and many-electron collective excitations (the so-called plasmons). The dielectric function of Equation (19.29) represents a very useful quantity for studying the excitation properties of the HEG. Specifically, from the proportionality of the doubly-differential inverse inelastic mean free path and the ELF (see Equation 19.42), it follows that the excitations of the materials (at any given q) should correspond to peaks in the $\mathrm{Im}[-1/\varepsilon(W,q)]$ vs. W plot. According now to Equation (19.35), the ELF has peaks at the maxima of the numerator and at the minima of the denominator. The first condition takes place when $\varepsilon_2(W,q)$ is maximum and $\varepsilon_1(W,q)$ remains almost constant, and corresponds to single-electron excitations. The second condition takes place when $\varepsilon_2(W,q) \ll 1$ and $\varepsilon_1(W,q) \approx 0$, so $|\varepsilon(W,q)| \approx 0$, which corresponds to many-electron collective excitations that are unique to the condensed phase since in the gas phase $|\varepsilon(W,q)| = 1$ always. These collective excitations emerge as follows. Whenever there is a sudden charge disturbance in the medium (e.g., by an incident charged particle), the charge carriers of the material will tend to rearrange (polarize) and screen off the disturbance owing to the long-range nature of the Coulomb interaction. Although screening would tend to diminish the strength of the particle-target interaction, as can be seen from Equation (19.33) under the condition $|\varepsilon(W,q)| > 1$, there appears also an opposite (i.e., antiscreening) effect due to the induced motion of the screening charges in the material leading to very small values of $|\varepsilon(W,q)|$. Specifically, as the charge carriers will tend to screen off the external charge, they will naturally overshoot their new equilibrium position, thus starting a collective oscillatory movement, a so-called *charge density oscillation*. The frequency of this oscillation can be shown to equal $\omega_{pl} = \sqrt{4\pi e^2 N_e/m}$. Thus, the energy quantum of this oscillation ($E_{pl} = \hbar\omega_{pl}$), the so-called plasmon, is given by Equation (19.32) or, in terms of r_s, by[*]

$$E_{pl} = \mathrm{Ry}(12/r_s^3)^{1/2} \qquad (19.59)$$

---

[*] For brevity, the dependence of E_{pl} upon r_s will be suppressed throughout; e.g., the notation $E_{pl}(r_s) \equiv E_{pl}$ will be implied.

EXAMPLE 19.2

Calculate the nominal plasmon energy of liquid water assuming the value of the electron gas parameter of liquid water is (see Example 19.1) $r_s = 1.69$.

SOLUTION 19.2

From Equation (19.59) we have

$$E_{pl} = 13.6\,\text{eV}\left(\frac{12}{1.69^3}\right)^{1/2} = 21.4\,\text{eV}$$

In many-body theory, the plasmon is the elementary excitation of this charge density oscillation, and is considered a quasi-particle consisting of the external charge along with its screening electronic cloud (Pines 1963). The defining condition of plasmon emerges as the solution of the following equation:

$$\varepsilon(W,q) = \varepsilon_1(W,q) + i\varepsilon_2(W,q) = 0 \Rightarrow \begin{cases} \varepsilon_1(E_{pl}(q),q) = 0 \\ \varepsilon_2(\Delta E_{pl}(q),q) = 0 \end{cases} \tag{19.60}$$

where $E_{pl}(q)$ and $\Delta E_{pl}(q)$ define the real and imaginary parts of the plasmon energy. The latter identifies with the plasmon line width or plasmon lifetime (see below). More specifically, $E_{pl}(q)$ and $\Delta E_{pl}(q)$ are the so-called *dispersion relations* that give the plasmon energy and line width, respectively, as a single-valued function of momentum transfer q. The conditions of Equation (19.60) lead to the following properties of the ELF in the plasmon region:

$$\text{Im}\left[\frac{-1}{\varepsilon(E_{pl}(q),q)}\right] \approx \text{Max}\left\{\text{Im}\left[\frac{-1}{\varepsilon(W,q)}\right]\right\} \tag{19.61}$$

and

$$\text{Im}\left[\frac{-1}{\varepsilon(E_{pl}(q) \pm \Delta E_{pl}(q),q)}\right] \approx \frac{1}{2}\text{Max}\left\{\text{Im}\left[\frac{-1}{\varepsilon(W,q)}\right]\right\} \tag{19.62}$$

The line width has an important physical interpretation. Specifically, via the energy-time uncertainty relation $\Delta E \approx \hbar/\tau$, the line width ($\Delta E_{pl}$) is inversely related to the lifetime (τ) of the excitation. Thus, an infinitesimal line width ($\Delta E_{pl} \approx 0$) implies a long-lived excitation lifetime ($\tau \approx \infty$). For the idealized case of a HEG we have $E_{pl}(q = 0) \equiv E_{pl}$, where E_{pl} is determined by Equation (19.32), and $\Delta E_{pl}(q = 0) = 0$, which implies an infinite plasmon lifetime. However, in real systems the plasmon resonance energy differs from the nominal plasmon energy E_{pl} calculated from Equation (19.32), while the plasmon peak has a finite line width ($\Delta E_{pl} \neq 0$) at all q and an associated finite lifetime. The plasmon decay rate (τ^{-1}) depends upon the efficiency of various energy-dissipative processes. Such processes include the excitation of multiple e-h pairs and the scattering of the excited electrons by impurities or ion cores.

It is important to recognize that properties of the ELF related to the position and width of its peaks (and their dispersion) will directly influence the calculated inelastic cross sections, which depend upon integrals of ELF over the energy-momentum plane (see Equations 19.46, 19.48, and 19.49). For example, other things being equal, the broader the ELF, the smaller the value of the inelastic mean free path at low incident energies (and vice versa) due to a larger fraction of ELF being within the integration limits. Thus, models of ELF that differ in their dispersion resulting in different plasmon energies (peak position) or lifetimes (peak width) are expected to result in, sometimes, sizable different inelastic cross sections.

The screening, as described by the dielectric function, has a direct effect upon the single-electron and plasmonic excitations of the material. First, let's point out that the magnitude of momentum transfer q, or equivalent, wavenumber $k = q/\hbar$, is indicative of the length scale of the perturbation induced in the material by the charged particle. For sufficiently large q (or k) the perturbation length scale will be restricted to single atoms, and one expects that the condensed phase effect will effectively vanish, i.e., $|\varepsilon(W,q)| \approx 1$, and the gas phase relation $\mathrm{Im}[-1/\varepsilon(W,q)] \approx \mathrm{Im}\,\varepsilon(W,q)$ is recovered. In the opposite limit of vanishing $q \approx 0$ (the so-called long wavelength or optical limit), the length scale can be large enough to encompass the entire system, and as a result, condensed phase effects are expected to be most pronounced in this limit. In this case, we distinguish among three excitation regimes: (1) For valence excitations with energy transfer much smaller than the plasmon energy we generally have $|\varepsilon(W,0)| > 1$ and, via Equation (19.35), $\mathrm{Im}[-1/\varepsilon(W,0)] < \mathrm{Im}\,\varepsilon(W,0)$. Thus, the effect of $\varepsilon(W,0)$ is to suppress the intensity of single-electron valence excitations, i.e., a typical *screening* effect. (2) For valence excitations at about the plasmon energy of the system ($W \approx E_{pl}$) we have $\varepsilon_1(E_{pl},0) \approx 0$, $\varepsilon_2(E_{pl},0) \ll 1$ and, as a result, $|\varepsilon(E_{pl},0)| \ll 1$. Then, from Equation (19.35) we obtain $\mathrm{Im}[-1/\varepsilon(E_{pl},0)] \gg \mathrm{Im}\,\varepsilon(E_{pl},0)$, which represents a sharp increase of the excitation probability. In general, the sharp increase exhibited by $\mathrm{Im}[-1/\varepsilon(E_{pl},0)]$ is larger than the maxima of $\mathrm{Im}\,\varepsilon(W,0)$. This represents a collective (plasmon-like) excitation of the valence electrons of the material; in analogy to the previous case we may call this an *antiscreening* effect. (3) For excitation energies much larger than the plasmon energy (i.e., in the energy range of inner shells) screening effects vanish since $|\varepsilon(W,0)| \approx 1$ and, consequently, $\mathrm{Im}[-1/\varepsilon(W,0)] \approx \mathrm{Im}\,\varepsilon(W,0)$, which is the typical gas phase, result. Thus, the condition $|\varepsilon(W,q)| \neq 1$, and as a corollary, $\mathrm{Im}[-1/\varepsilon(W,q)] \neq \varepsilon_2(W,q)$, which signifies that condensed phase effects are present, is realized mainly at $q \to 0$ and at values of W in the range of valence electron excitations. To summarize:

- Screening ($W \ll E_{pl}$ & $q \approx 0$):

$$|\varepsilon(W,0)| > 1 \Rightarrow \mathrm{Im}\left[\frac{-1}{\varepsilon(W,0)}\right] < \mathrm{Im}\,\varepsilon(W,0) \tag{19.63}$$

- Antiscreening ($W \approx E_{pl}$ & $q \approx 0$):

$$\left.\begin{array}{l} \varepsilon_1(E_{pl},0) \approx 0 \\[2mm] \varepsilon_2(E_{pl},0) \ll 1 \end{array}\right\} \Rightarrow |\varepsilon(E_{pl},0)| \ll 1 \Rightarrow \mathrm{Im}\left[\frac{-1}{\varepsilon(E_{pl},0)}\right] \gg \mathrm{Im}\,\varepsilon(E_{pl},0) \tag{19.64}$$

- No screening ($W \gg E_{pl}$ & *any q*):

$$|\varepsilon(W,q)| \approx 1 \Rightarrow \mathrm{Im}\left[\frac{-1}{\varepsilon(W,q)}\right] \approx \mathrm{Im}\,\varepsilon(W,q) \tag{19.65}$$

19.5.2 Drude Model

The Drude model provides a very simple yet useful approximation to the dielectric response of condensed media. It is assumed that under a spatially uniform electric field of frequency $\omega = E/\hbar$, the screening electrons are set in a harmonic oscillatory motion of frequency equal to the plasmon frequency of the material, $\omega_{pl} = E_{pl}/\hbar$, which is "damped" by a friction-like force due to general dissipative processes that are represented by a finite damping constant ($\gamma \neq 0$). The latter is inversely proportional to the plasmon lifetime τ via the relation $\gamma = \hbar/\tau$. The Drude dielectric function reads:

$$\varepsilon_D(W;E_{pl},\gamma) = 1 - \frac{E_{pl}^2}{W^2 + iW\gamma} \tag{19.66}$$

with its real and imaginary parts being, respectively:

$$\mathrm{Re}\,\varepsilon_D(W;E_{pl},\gamma) = 1 - \frac{E_{pl}^2}{W^2 + \gamma^2} \tag{19.67}$$

and

$$\mathrm{Im}\,\varepsilon_D(W;E_{pl},\gamma) = \frac{\gamma E_{pl}^2}{W(W^2 + \gamma^2)} \tag{19.68}$$

From the plasmon condition, Equation (19.60), and setting $\varepsilon(E_{res}, q = 0) = \varepsilon_D(E_{res}; E_{pl}, \gamma)$, we obtain the solution

$$\varepsilon_D(E_{res}; E_{pl}, \gamma) = 0 \Rightarrow \begin{cases} \text{Re}\,\varepsilon_D(E_{res}; E_{pl}, \gamma) = 0 \Rightarrow E_{res} \approx (E_{pl}^2 - \gamma^2)^{1/2} \\ \text{Im}\,\varepsilon_D(\Delta E_{res}; E_{pl}, \gamma) = 0 \Rightarrow \Delta E_{res} \approx \gamma/2 \end{cases} \quad (19.69)$$

It follows that the damping constant γ, which determines the line width ΔE_{res}, causes the resonance energy E_{res} to be displaced to *smaller* values from the nominal plasmon value E_{pl}.

The ELF obtained from the Drude dielectric function, Equation (19.66), reads:

$$\text{Im}\left[\frac{-1}{\varepsilon_D(W; E_{pl}, \gamma)}\right] = \frac{E_{pl}^2 \gamma W}{\left(E_{pl}^2 - W^2\right)^2 + (\gamma W)^2} \quad (19.70)$$

Thus, the Drude-ELF is a Lorentz-like function with a *full-width-at-half-maximum* (FWHM) given by

$$\Delta E_{1/2} \approx \gamma \quad (19.71)$$

and a maximum (i.e., peak height) that, for *small* γ, is located at about the plasmon energy E_{pl} and amounts to

$$\text{Max}\left\{\text{Im}\left[\frac{-1}{\varepsilon_D(W; E_{pl}, \gamma)}\right]\right\} \approx \frac{E_{pl}}{\gamma} \quad (19.72)$$

In Figure 19.2 we depict the real (ε_1) and imaginary (ε_2) parts of the Drude dielectric function as well as its ELF ($-\text{Im}\,\varepsilon^{-1}$) calculated, respectively, from Equations (19.67), (19.68), and (19.70) for a material with $E_{pl} = 20$ eV (or $r_s = 1.77$) and for two different values of the damping constant ($\gamma = 0.1$ eV and $\gamma = 1$ eV). We may first notice that, as follows from Equation (19.69), $E_{res} \approx E_{pl}$ since $\gamma \ll E_{pl}$ for both cases. Also, with the increase of the damping constant, the width of ELF increases, whereas its peak height decreases in accordance, respectively, to Equations (19.71) and (19.72). At the inset we provide a close-up view of the excitation region around the plasmon resonance where ε_1 crosses the horizontal axis (becoming zero at $W \equiv E_{res} \approx E_{pl}$) and $\varepsilon_2 \ll 1$, so $\text{Im}(-1/\varepsilon) \approx 1/\varepsilon_2$ becomes maximum. At higher energies (not shown) $\varepsilon_1 \to 1$, $\varepsilon_2 \to 0$, and $|\varepsilon| \approx 1$, which is the condition described in Equation (19.65) where screening vanishes and we recover the gas phase condition.

It can be shown that the *f*-sum rule is exactly satisfied by the Drude dielectric function since

$$\int_0^\infty W \,\text{Im}\,\varepsilon_D(W; E_{pl}, \gamma)\,dW = \int_0^\infty W \,\text{Im}\left[\frac{-1}{\varepsilon_D(W; E_{pl}, \gamma)}\right]dW = \frac{\pi}{2} E_{pl}^2 \quad (19.73)$$

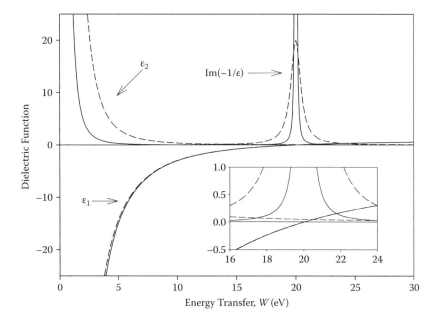

FIGURE 19.2

The Drude dielectric function for two different values of the damping constant (full line: $\gamma = 0.1$ eV; broken line: $\gamma = 1$ eV). The inset provides a close-up view of the plasmon resonance region. Calculations are based on typical values for organic matter: $E_{pl} = 20$ eV or $r_s = 1.77$.

In Figure 19.3 we have plotted separately the f-sum for Im $\varepsilon_D(W; E_{pl}, \gamma)$ and Im$[-1/\varepsilon_D(W; E_{pl}, \gamma)]$ (both multiplied by $2/\pi E_{pl}^2$, i.e., normalized to unity) as a function of the upper integration limit of Equation (19.73) for the model parameters used in Figure 19.2. It can be seen that there is an abrupt increase of the f-sum for the ELF in the excitation region of the plasmon peak. Note also that within the Drude model the plasmon fully exhausts the f-sum rule. As will be shown in the next section, this is also true in the quantum mechanical treatment of the HEG within the so-called random phase approximation (RPA), but only at $q = 0$.

It follows from Equation (19.70) that for $\gamma \ll E_{pl}$ the Drude-ELF exhibits a plasmon peak of infinitesimal width and large height. Physically this represents a condition whereby energy-dissipative processes are absent and the plasmon is a long-lived excitation ($\tau \rightarrow \infty$) or, in the language of many-body theory, a well-defined quasi-particle. Using the properties of the delta-function, the ELF for an *undamped* ($\gamma \approx 0$) Drude model can be obtained from Equation (19.70) and reads:

$$\lim_{\gamma \rightarrow 0} \text{Im}\left[\frac{-1}{\varepsilon_D(W; E_{pl}, \gamma)} \right] = \frac{\pi}{2} E_{pl} \delta(W - E_{pl}) \qquad (19.74)$$

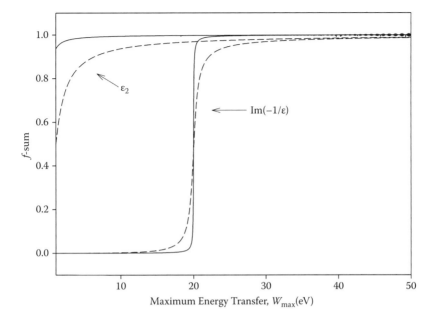

FIGURE 19.3
The *f*-sum for the imaginary part of the Drude dielectric function and energy loss function as a function of the maximum energy transfer (model parameters are the same as in Figure 19.2).

Evidently, the maximum of the undamped Drude-ELF appears at exactly $W = E_{pl}$. It is trivial to show that the undamped Drude model also satisfies the *f*-sum rule:

$$\int_0^\infty W \left\{ \lim_{\gamma \to 0} \mathrm{Im} \left[\frac{-1}{\varepsilon_D(W; E_{pl}, \gamma)} \right] \right\} dW = \frac{\pi}{2} E_{pl} \int_0^\infty W\, \delta(W - E_{pl})\, dW = \frac{\pi}{2} E_{pl}^2 \quad (19.75)$$

It is apparent from the above discussion that the original Drude model is useful for describing the collective electronic excitations (i.e., plasmons) of condensed matter. A modified form of the Drude model can be used to also account for interband single-electron excitations (i.e., e-h pair excitations). The *modified Drude* (MD) dielectric function appropriate for interband excitations reads:

$$\varepsilon_{MD}(W; E_{pl}, f_n, E_n, \gamma_n) = 1 + \sum_n f_n \frac{E_{pl}^2}{E_n^2 - W^2 - iW\gamma_n} \quad (19.76)$$

where f_n, E_n, and γ_n represent, respectively, the oscillator strength, excitaton energy, and damping constant of the nth interband excitation. As expected, setting $f_n = 1$, $E_n = 0$, and $\gamma_n - \gamma$ in Equation (19.76), we recover the original

Drude dielectric function of Equation (19.66), since, in the case of the HEG, the plasmon exhausts the sum rule (at $q = 0$); i.e., the probability for interband excitations is zero.

The real and imaginary parts of the modified Drude dielectric function, Equation (19.76), are

$$\operatorname{Re}\varepsilon_{\mathrm{MD}}(W;E_{pl},f_n,E_n,\gamma_n)=1+E_{pl}^2\sum_n f_n\frac{(E_n^2-W^2)}{(E_n^2-W^2)^2+(W\gamma_n)^2} \quad (19.77)$$

$$\operatorname{Im}\varepsilon_{\mathrm{MD}}(W;E_{pl},f_n,E_n,\gamma_n)=E_{pl}^2\sum_n f_n\frac{W\gamma_n}{(E_n^2-W^2)^2+(W\gamma_n)^2} \quad (19.78)$$

and the ELF is obtained in the standard way from

$$\operatorname{Im}\left[\frac{-1}{\varepsilon_{\mathrm{MD}}(W;E_{pl},f_n,E_n,\gamma_n)}\right]$$

$$=\frac{\operatorname{Im}\varepsilon_{\mathrm{MD}}(W;E_{pl},f_n,E_n,\gamma_n)}{[\operatorname{Re}\varepsilon_{\mathrm{MD}}(W;E_{pl},f_n,E_n,\gamma_n)]^2+[\operatorname{Im}\varepsilon_{\mathrm{MD}}(W;E_{pl},f_n,E_n,\gamma_n)]^2} \quad (19.79)$$

The f-sum rule for the modified Drude function reads:

$$\int_0^\infty W\operatorname{Im}\varepsilon_{\mathrm{MD}}(W;E_{pl},f_n,E_n,\gamma_n)dW=\frac{\pi}{2}E_{pl}^2\sum_n f_n \quad (19.80)$$

which is satisfied as long as $\sum_n f_n = 1$ independent of the values of the coefficients E_n and γ_n. Note that if the f-sum rule is satisfied by $\operatorname{Im}\varepsilon_{\mathrm{MD}}(W;E_{pl},f_n,E_n,\gamma_n)$, it will then be also satisfied by $\operatorname{Im}[-1/\varepsilon_{\mathrm{MD}}(W;E_{pl},f_n,E_n,\gamma_n)]$. A general limitation of the Drude model is that the q-dependence of the dielectric function is not considered; i.e., the Drude dielectric function is *local* in the sense that the variation of the plasmon oscillation (or interband excitation) in space is neglected.

19.5.3 Lindhard Model

To go beyond the Drude model one has to treat the screening quantum mechanically. It is important to highlight that even for such a simple model as the HEG an exact expression for its dielectric function is not available. The reason is that although in the HEG the (screening) electrons do not interact with the ion cores (they are considered to be unbound), they can interact with each other (i.e., e-e collisions are present). Specifically, due to Coulomb repulsive forces and the Pauli exclusion principle, there is a reduced electronic density around each electron that produces the so-called Coulomb and exchange holes. The correlation in the motion of electrons due to e-e interactions (the famous many-body problem in condensed matter physics) is what makes the dielectric function of the HEG intractable. A solution to the above problem is provided by making yet another drastic approximation to the HEG, the so-called *random phase approximation* (RPA), which assumes that each electron interacts with the *average* field generated by all other electrons. In essence, this is equivalent to accounting only for *long-range correlation* in electron motion (equivalent to a time-dependent Hartree approximation). Hence, RPA is "exact" only at the limit of infinite electronic density ($r_s \rightarrow 0$), where *short-range correlation* is suppressed. To emphasize the neglect of short-range correlation in e-e interactions, the RPA is often said to describe the *noninteracting* HEG despite the fact that e-e interactions are considered, but only in an average manner.

The dielectric function of the HEG within the RPA for arbitrary values of momentum transfer q was first derived by Lindhard (1954) and reads:

$$\varepsilon_L(W, q; E_{pl}) = 1 + (\chi^2/z^2) f(x, z) \tag{19.81}$$

where

$$f(x, z) = \frac{1}{2} + \frac{1}{8z}\{1 - (z - x/4z)^2\}\ln\left|\frac{z - x/4z + 1}{z - x/4z - 1}\right|$$

$$+ \frac{1}{8z}\{1 - (z + x/4z)^2\}\ln\left|\frac{z + x/4z + 1}{z + x/4z - 1}\right| \tag{19.82}$$

with $z = q/2q_F$, $x = (W + i\gamma)/E_F$, and $\chi^2 = e^2/\pi\hbar\upsilon_F$. The Fermi momentum ($q_F$) and energy ($E_F$) can be conveniently expressed in terms of the electron gas parameter r_s via the relation

$$E_F = \frac{Ry}{(r_s\alpha)^2} \tag{19.83}$$

with $\alpha = (\frac{4}{9\pi})^{1/3} = 0.521$.

EXAMPLE 19.3

Calculate the nominal Fermi energy of liquid water assuming the value of the electron gas parameter of liquid water is (see Example 19.1) $r_s = 1.69$.

SOLUTION 19.3

From Equation (19.83) we have

$$E_F = \frac{13.6\,\text{eV}}{(0.521 \times 1.69)^2} = 17.5\,\text{eV}$$

Since in the RPA dissipative processes are neglected, we should take the limit of zero plasmon damping ($\gamma \to 0$), or the equivalent, of infinite plasmon lifetime ($\tau \to \infty$) in the Lindhard dielectric function. In this collisionless limit the Lindhard dielectric function reads:

$$\varepsilon_L(W, q; E_{pl}) = 1 + (\chi^2/z^2)[f_1(x, z) + if_2(x, z)] \qquad (19.84)$$

with the real and imaginary parts, respectively, being $\mathrm{Re}\, \varepsilon_L(W, q; E_{pl}) = 1 + (\chi^2/z^2)\, f_1(x, z)$ and $\mathrm{Im}\,\varepsilon_L(W, q; E_{pl}) = (\chi^2/z^2) f_2(x, z)$. The expressions for $f_1(x, z)$ and $f_2(x, z)$ read:

$$f_1(x, z) = \frac{1}{2} + \frac{1}{8z} \{1 - (z - x/4z)^2\} \ln \left| \frac{z - x/4z + 1}{z - x/4z - 1} \right|$$

$$+ \frac{1}{8z} \{1 - (z + x/4z)^2\} \ln \left| \frac{z + x/4z + 1}{z + x/4z - 1} \right| \qquad (19.85)$$

and

$$f_2(x, z) = \begin{cases} \pi x/8z, & 0 < x < 4z(1 - z) \\ (\pi/8z)\{1 - (z - x/4z)^2\}, & 4z(z - 1) < x < 4z(z + 1) \quad \text{and} \quad x > 4z(1 - z) \\ \approx 0, & \text{otherwise} \end{cases}$$

$$(19.86)$$

where now $x = W/E_F$. In Figure 19.4 we present the real and imaginary parts of the (collisionless) Lindhard dielectric function as well as the corresponding ELF calculated from Equations (19.84) to (19.86) for three different values of momentum transfer q. The arrows in the ELF plot (Figure 19.4c) indicate the position of the plasmon peak, which, in the RPA, has zero width and infinite height. Contrary to the Drude model, the imaginary part is sizable over an appreciable range of energy transfers due to the presence of single-electron excitations (i.e., e-h pairs). The contribution of single-electron excitations is also clearly seen in the ELF plot. The contribution of single-electron and plasmon excitations in different regions of the W-q plane can be deduced by inspection of $f_2(x, z)$ and, correspondingly, of $\mathrm{Im}\,\varepsilon_L(W, q; E_{pl})$. The form of Equation (19.86) suggests that we can distinguish between three regions in the x-z (or W-q) plane as shown in Figure 19.5.

- **Region I:** Defined by the condition $0 < x < 4z(1 - z)$. In this region $x \leq 1$ or $W \leq E_F$, so single-electron excitations cannot be excited. Also, since $f_2^I(x, z) = \pi x/8z$ is finite and, as a result, $\mathrm{Im}\,\varepsilon_L(W, q; E_{pl}) \neq 0$, neither plasmon excitation is possible in this region.

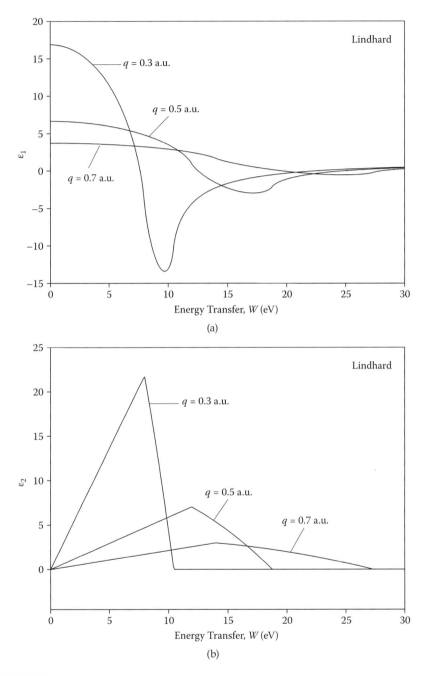

FIGURE 19.4
The Lindhard dielectric function (a and b) and energy loss function (c) at different values of momentum transfer. Calculations are based on typical values for organic matter: $E_{pl} = 20$ eV or $r_s - 1.77$. The arrows in (c) indicate the position of the plasmon peaks.

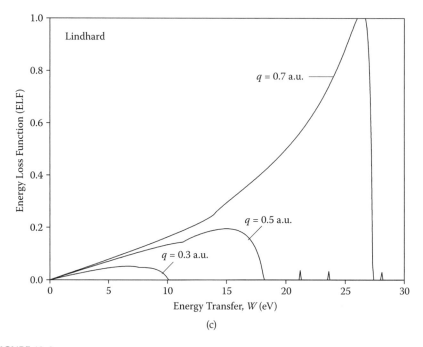

FIGURE 19.4
(Continued)

- **Region II:** Defined by the conditions $4z(z - 1) < x < 4z(z + 1)$ and $x > 4z(1 - z)$. In this region $f_2^{II}(x,z) = (\pi/8z)\{1 - (z - x/4z)^2\}$ and $\mathrm{Im}\,\varepsilon_L(W,q;E_{pl}) \neq 0$, and single-electron excitations are dominant.

- **Region III:** Consists of two separate regions, one between the lines $z = 0$ and $x = 4z(z + 1)$ and the other defined by the condition $0 < x < 4z(z - 1)$. In this region we have $f_2^{III}(x,z) \approx 0$ and $\mathrm{Im}\,\varepsilon_L(W,q;E_{pl}) \approx 0$. Thus, single-electron excitations are not possible but plasmon excitations arise along the dispersion line defined by the condition $\mathrm{Re}\,\varepsilon_L(W,q;E_{pl}) \approx \varepsilon_L(W,q;E_{pl}) = 0$.

The plasmon dispersion line $E_{pl}(q)$ can be found by solving $\varepsilon_L(W,q;E_{pl}) = 0$ numerically. However, an approximate *analytic* solution is possible at small q by considering $\varepsilon_L(W,q;E_{pl})$ for $W \approx E_{pl}$. A low-q expansion of $\varepsilon_L(W,q;E_{pl})$ in even powers of q reads:

$$\varepsilon_L(W,q;E_{pl}) \approx 1 - \frac{E_{pl}^2}{W^2} - \frac{6}{5}\frac{E_{pl}^2 E_F}{W^4}\frac{q^2}{m} + \cdots \tag{19.87}$$

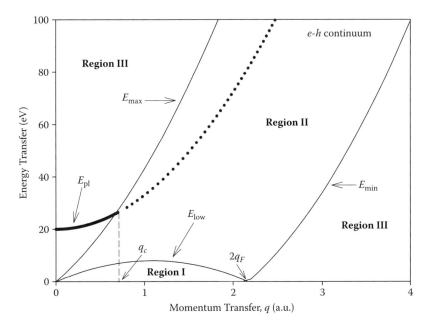

FIGURE 19.5
Regions in the energy-momentum plane where different excitations of the HEG are possible. $E_{min} = q^2/2m - qv_F$, $E_{max} = q^2/2m + qv_F$, and $E_{low} = qv_F - q^2/2m$. Calculations are based on typical values for organic matter: $E_{pl} = 20\,\text{eV}$ or $r_s = 1.77$.

Then, using the defining condition of plasmon dispersion $\varepsilon_L(W, q; E_{pl}) = 0$ we obtain for small q and up to second order in q:

$$E_{pl}^{RPA}(q) = E_{pl} + \alpha_{RPA}\frac{q^2}{2m} \tag{19.88}$$

where $E_{pl} \equiv E_{pl}(q=0)$ and the dimensionless dispersion coefficient $\alpha_{RPA} \equiv \alpha_{RPA}(r_s)$ is given by[*]

$$\alpha_{RPA} = \frac{6}{5}\frac{E_F}{E_{pl}} \tag{19.89}$$

---

[*] Sometimes the RPA dispersion coefficient is defined as $\alpha_{RPA}=(3/5)(E_F/E_{pl})$. Then, the second term in the right-hand side of Equation (19.88) must be changed accordingly from $\alpha_{RPA}(q^2/2m)$ to $\alpha_{RPA}(q^2/m)$.

EXAMPLE 19.4

Calculate the RPA quadratic dispersion coefficient of liquid water.

SOLUTION 19.4

From Examples 19.2 and 19.3 and Equation (19.89) we have

$$\alpha_{RPA} = \frac{6}{5} \times \frac{17.5\,eV}{21.4\,eV} = 0.981$$

From Equation (19.87) it follows that at small q and $W \approx E_{pl}$ the Lindhard dielectric function is real; therefore:

$$\mathrm{Im}\,\varepsilon_L(W \approx E_{pl}, q; E_{pl}) = 0 \Rightarrow \Delta E_{pl}^{RPA}(q) = 0 \tag{19.90}$$

In fact, it can be shown that at the optical limit ($q = 0$) the Lindhard-ELF reduces to the *undamped* Drude-ELF:

$$\lim_{q \to 0} \mathrm{Im}\left[\frac{-1}{\varepsilon_L(W, q; E_{pl})}\right] = \lim_{\gamma \to 0} \mathrm{Im}\left[\frac{-1}{\varepsilon_D(W; E_{pl}, \gamma)}\right] = \frac{\pi}{2} E_{pl}\delta(W - E_{pl}) \tag{19.91}$$

However, contrary to the Drude dielectric function, the Lindhard dielectric function provides automatically the extension to $q \neq 0$ (i.e., it is *nonlocal*) with the *f*-sum rule fulfilled at any q. According to Equation (19.91) the plasmon is the sole excitation in the HEG (within the RPA) at $q = 0$; i.e., it fully exhausts the *f*-sum rule as already discussed in Figure 19.3.

19.5.4 Landau Damping

Although according to Equation (19.90) plasmons are undamped at small q, the Lindhard dielectric function *does* predict plasmon damping, the so-called *Landau damping*, above a critical momentum transfer value q_c, where for $q > q_c$ and $W \approx E_{pl}$ the imaginary part of the Lindhard dielectric function becomes finite, i.e., $\mathrm{Im}\,\varepsilon_L(W, q; E_{pl})_{q > q_c} \neq 0$. A rough estimate of the critical momentum transfer $q_c = \hbar k_c$ can be obtained from the relation[*]

$$k_c \approx \frac{\omega_{pl}}{\upsilon_F} \tag{19.92}$$

EXAMPLE 19.5

Calculate the critical momentum transfer for liquid water in both atomic units and angstroms.

---

[*] For a more exact expression, see, for example, Egerton (1996, p. 161).

SOLUTION **19.5**

In atomic units $\hbar = 1$ so $\omega_{pl} = E_{pl}$. Then, from Example 19.2, we have

$$\omega_{pl} = \frac{21.4 \text{ eV}}{27.2 \text{ eV}} = 0.787 \text{ a.u.}$$

Also, in atomic units $m = 1$ so $\upsilon_F = (2E_F)^{1/2}$. Then, from Example 19.3 we have

$$\upsilon_F = \left(2 \times \frac{17.5 \text{ eV}}{27.2 \text{ eV}} \right)^{1/2} = 1.13 \text{ a.u.}$$

Substituting to Equation (19.92) we obtain:

$$k_c = \frac{0.787}{1.13} = 0.696 \text{ a.u.}$$

or

$$k_c = \frac{0.696}{0.529 \text{ Å}} = 1.32 \text{ Å}^{-1}$$

The explanation of the Landau damping mechanism is the following: for $q > q_c$ the plasmon energy calculated from the plasmon dispersion relation $E_{pl}(q)$ enters Region II of the e-h pair excitation continuum, i.e., $q^2/2m - q\upsilon_F < E_{pl}(q) < q^2/2m + q\upsilon_F$ (see Figure 19.5). Then, as a result of energy momentum considerations, the plasmon can decay by transferring all its energy to a *single* electron, which can then undergo an intraband excitation generating an e-h pair.

In Figure 19.6 we depict the Lindhard-ELF for three values of the momentum transfer above the Landau critical value where the d-like plasmon peak has transformed to a broad resonance due to the decay of plasmons to *single* e-h pair excitations. It can be seen that, in this case, the corresponding plasmon energies (denoted by filled triangles in the energy transfer axis) calculated by the RPA plasmon dispersion relation of Equation (19.88) fall within the range of single-electron excitations. The latter provide a highly efficient dissipation mechanism; therefore, the plasmon is now so strongly damped (i.e., $\tau \to 0$) that the concept of plasmon itself as a well-defined quasi-particle becomes meaningless. The strong damping of plasmons by the Landau mechanism when they enter the single-electron continuum is better visualized in Figure 19.7, where the ELF computed by the Lindhard dielectric function is depicted as both a 3D surface and a 2D contour plot.

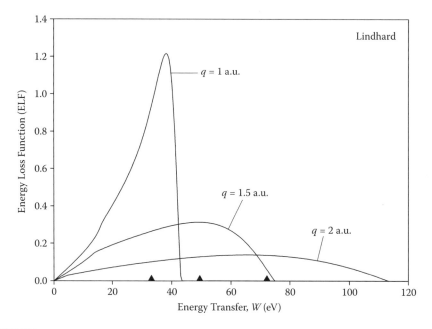

FIGURE 19.6

The Lindhard energy loss function at momentum transfer values above the Landau cutoff. The filled triangles along the energy transfer axis denote the position of the corresponding plasmon energies. Calculations are based on typical values for organic matter: $E_{pl} = 20$ eV or $r_s = 1.77$.

19.5.5 Mermin Model

One of the shortcomings of the Lindhard dielectric function is the neglect of the finite lifetime of plasmons at small q ($q < q_c$), which gives rise to a finite line width of the plasmon peak in the ELF. The prediction of an undamped plasmon is an oversimplification since, in real materials, plasmons exhibit a finite lifetime at any q due to their decay by two (or more) e-h pairs, e-ion collisions, or e-scattering by imperfections (e.g., impurities). The inclusion of damping in the Lindhard dielectric function, Equation (19.84), through a relaxation time approximation whereby the excitation energy W is simply replaced by a complex excitation energy $\tilde{W} = W + i\gamma$ is known to be incorrect because it does not conserve the local number of particles (Kliewer and Fuchs 1969). The correct expression for the dielectric function of the HEG within the RPA that includes a *phenomenological* damping constant that represents phonon-assisted electronic transitions via e-ion collisions was proposed by Mermin (1970). The Mermin dielectric function is obtained by an appropriate combination of terms that involve the Lindhard dielectric function of complex energy ($\tilde{W}$) as follows:

$$\varepsilon_M(W,q;E_{pl},\gamma) = 1 + \frac{(1+i\gamma/W)[\varepsilon_L(\tilde{W},q;E_{pl})-1]}{1+(i\gamma/W)[\varepsilon_L(\tilde{W},q;E_{pl})-1]/[\varepsilon_L(W=0,q;E_{pl})-1]} \quad (19.93)$$

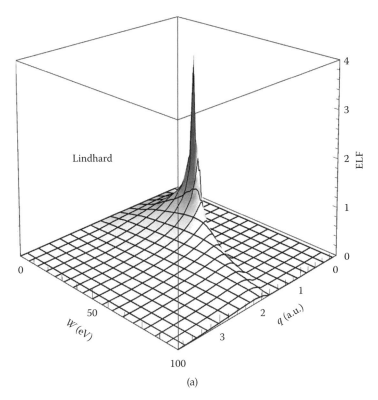

(a)

FIGURE 19.7
The Lindhard energy loss function as a function of energy transfer W and momentum transfer q depicted as a 3D surface (a) and a 2D contour (b). Calculations are based on typical values for organic matter: $E_{pl} = 20$ eV or $r_s = 1.77$.

where γ is the phenomenological damping constant, the value of which, as in the Drude model, is not specified within the model and has to be estimated by other means. It can be easily verified that at the limit of zero damping the Mermin dielectric function recovers the Lindhard dielectric function:

$$\lim_{\gamma \to 0} \varepsilon_M(W, q; E_{pl}, \gamma) = \varepsilon_L(W, q; E_{pl}) \qquad (19.94)$$

Thus, the Mermin dielectric function also satisfies the f-sum rule for any q.

In Figure 19.8 we present an ELF (at a fixed momentum transfer, $q = 1$ a.u.) as a function of energy transfer for three different values of γ. Evidently with increasing γ the width of the ELF increases, resulting in a more diffused plasmon peak and the emergence of a high-energy-loss tail. The Bethe surface of the ELF computed by the Mermin dielectric function (with $\gamma = 10$ eV) is depicted in Figure 19.9. The effect of a finite damping constant ($\gamma \neq 0$) that enters into the Mermin dielectric function can be seen in Figure 19.10, where we present 2D contour plots of the ELF as a function

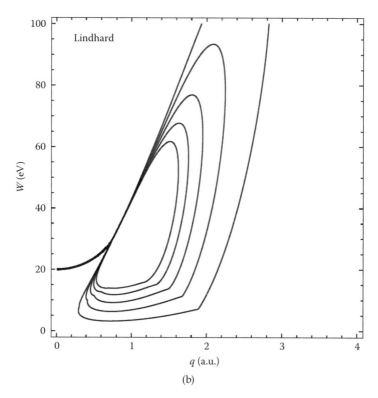

FIGURE 19.7
(Continued)

of energy and momentum transfer for three different values of γ. For small values of γ the effect of the Landau damping mechanism is still evident above a certain momentum transfer ($q > q_c$). However, with increasing values of γ there is a strong damping even at smaller momentum transfer (i.e., below the onset of the Landau damping mechanism, $q < q_c$).

In Figure 19.11 we compare the ELFs computed by the Lindhard and Mermin (with γ = 5 eV) dielectric functions for three different values of q. For the lowest q depicted, which is smaller than the Landau cutoff, the Lindhard-ELF exhibits a well-separated single-electron excitation region and a d-like plasmon peak. The Mermin-ELF exhibits a more diffused plasmon peak than the Lindhard-ELF for all q, with differences gradually vanishing at large q. It can be shown that the Mermin-ELF reduces to the *damped* Drude-ELF at the optical limit:

$$\lim_{q \to 0} \mathrm{Im}\left[\frac{-1}{\varepsilon_M(W,q;E_{pl},\gamma)} \right] = \mathrm{Im}\left[\frac{-1}{\varepsilon_D(W;E_{pl},\gamma)} \right] = \frac{E_{pl}^2\, \gamma\, W}{\left(E_{pl}^2 - W^2\right)^2 + (\gamma W)^2} \qquad (19.95)$$

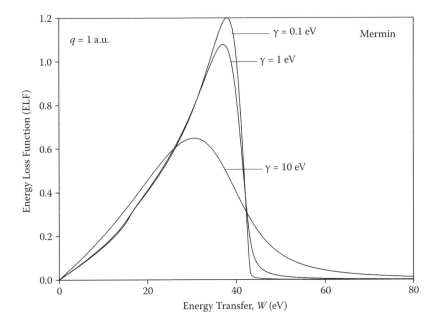

FIGURE 19.8
The Mermin energy loss function at a fixed momentum transfer ($q = 1$ a.u.) for different values of the damping constant γ. Calculations are based on typical values for organic matter: $E_{pl} = 20$ eV or $r_s = 1.77$.

19.5.6 Plasmon Pole Approximation

A widely used simplification of the RPA dielectric function is obtained under the so-called *plasmon pole* (PP) approximation first suggested by Hedin and Lundqvist (1969). In the PP approximation one assumes that the excitation spectrum of the HEG can be replaced by a single pole along the plasmon dispersion line, $E_{pl}(q)$. Importantly, the form of the dielectric function in the PP approximation makes it possible to account for the finite plasmon lifetime through the inclusion of a damping constant, thus making it roughly equivalent to the Mermin dielectric function. Furthermore, the damping constant can be made q-dependent via an appropriate dispersion relation. In the general case of a finite and q-dependent damping constant, the dielectric function in the PP approximation reads:

$$\varepsilon_{PP}(W; E_{pl}(q), \gamma(q)) = 1 + \frac{E_{pl}^2}{E_{pl}^2(q) - E_{pl}^2 - W^2 - iW\gamma(q)} \tag{19.96}$$

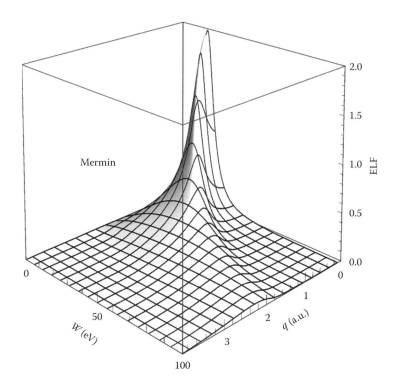

FIGURE 19.9
The Mermin energy loss function (with $\gamma = 1$ eV) as a function of energy transfer W and momentum transfer q depicted as a 3D surface. Calculations are based on typical values for organic matter: $E_{pl} = 20$ eV or $r_s = 1.77$.

with the corresponding ELF being

$$\text{Im}\left[\frac{-1}{\varepsilon_{PP}(W;E_{pl}(q),\gamma(q))}\right] = \frac{E_{pl}^2 \, \gamma(q) W}{\left[E_{pl}^2(q) - W^2\right]^2 + [W\gamma(q)]^2} \qquad (19.97)$$

The form of Equation (19.97) leads to the following conclusion: the ELF in the PP approximation reduces to the Drude-ELF of Equation (19.70) at the optical limit ($q = 0$):

$$\lim_{q \to 0} \text{Im}\left[\frac{-1}{\varepsilon_{PP}(W;E_{pl}(q),\gamma(q))}\right] = \text{Im}\left[\frac{-1}{\varepsilon_D(W;E_{pl},\gamma)}\right] \qquad (19.98)$$

while its extension to $q \neq 0$ is obtained after the replacement: $E_{pl} \to E_{pl}(q)$ and $\gamma \to \gamma(q)$. Therefore, the PP approximation is also referred to in the literature as an extended Drude model.

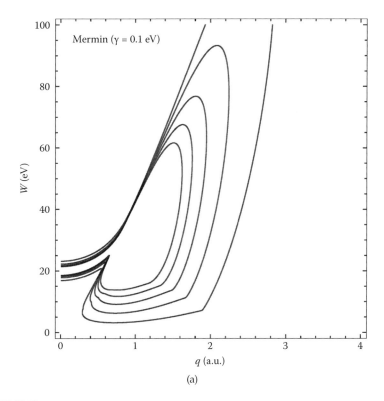

(a)

FIGURE 19.10

The Mermin energy loss function as a function of energy transfer W and momentum transfer q depicted as a 2D contour plot for three different values of the damping constant: $\gamma = 0.1$ eV (panel a), $\gamma = 1$ eV (panel b), $\gamma = 10$ eV (panel c). Calculations are based on typical values for organic matter: $E_{pl} = 20$ eV or $r_s = 1.77$.

In the limit of zero damping ($\gamma \approx 0$) whereby the complete excitation spectrum is "collapsed" to a single peak of zero width along the plasmon dispersion line we obtain:[*]

$$\lim_{\gamma(q)\to 0} \mathrm{Im}\left[\frac{-1}{\varepsilon_{PP}(W;E_{pl}(q),\gamma(q))} \right] = \frac{\pi}{2} \frac{E_{pl}^2}{E_{pl}(q)} \delta(W - E_{pl}(q)) \qquad (19.99)$$

At the optical limit Equation (19.99) reduces to the ELF of the *undamped* Drude model of Equation (19.74).

It is important to note that from the properties of the Drude function discussed in Section 19.5.2, it follows that the PP approximation satisfies the f-sum rule irrespective of the form of the dispersion relation for $E_{pl}(q)$ and $\gamma(q)$.

---

[*] In the limit of zero damping, the plasmon pole approximation is often called one-mode or single-pole approximation.

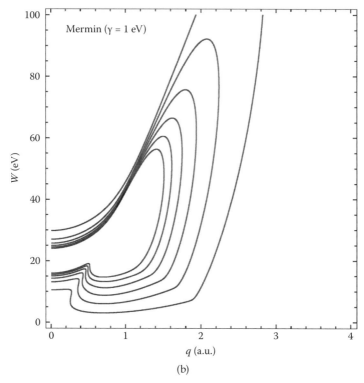

FIGURE 19.10
(Continued)

19.5.7 Many-Body Local Field Correction

To go beyond RPA one needs to consider the so-called *many-body local field correction* (LFC),[*] which is meant to account for short-range correlation (due to Coulomb repulsion) and exchange (due to the Pauli principle) that are neglected in the RPA and which result in a local depletion of electron density around each electron, giving rise to the so-called exchange-correlation (XC) hole. In the context of the LFC theory, one speaks of an *interacting* HEG to emphasize the inclusion of e-e interactions beyond the RPA, and the term *electron liquid* often replaces that of *electron gas*.

As first shown by Hubbard (1958), the dielectric function of the *interacting* HEG (or electron liquid) can be written as follows:[†]

$$\varepsilon_{\text{HEG}}(W,q) = 1 + \frac{\chi_{\text{RPA}}(W,q)}{1 - G(W,q)\chi_{\text{RPA}}(W,q)} \qquad (19.100)$$

---

[*] Not to be confused with the *crystal* LFC arising from the inhomogeneous nature of electron density due to band-structure effects.

[†] Also called the test charge–test charge dielectric function since it determines the screening provided by an electron gas to two "test" charges.

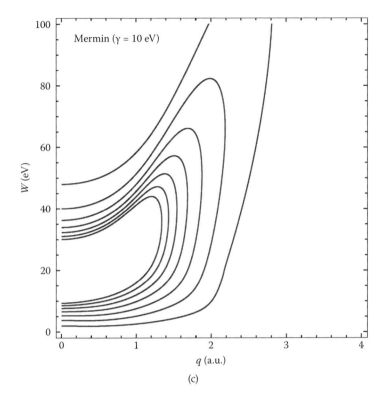

(c)

FIGURE 19.10
(Continued)

with $\chi_{RPA}(W,q) \equiv \varepsilon_L(W,q) - 1$ being the complex electron susceptibility in the RPA and $G(W,q) = G_1(W,q) + iG_2(W,q)$ being the LFC, which, in general, is a complex-valued function depending on both the energy and momentum transfer. Then, the ELF of the *interacting* HEG reads:

$$\text{Im}\left[\frac{-1}{\varepsilon_{HEG}(W,q)}\right] = \text{Im}\left\{\frac{\chi_{RPA}(W,q)}{1 + [1 - G(W,q)]\chi_{RPA}(W,q)}\right\} \quad (19.101)$$

or

$$\text{Im}\left[\frac{-1}{\varepsilon_{HEG}(W,q)}\right] = \frac{\chi_2^{RPA} + G_2\left[\left(\chi_1^{RPA}\right)^2 + \left(\chi_2^{RPA}\right)^2\right]}{\left(1 + \chi_1^{RPA} - G_1\chi_1^{RPA} + G_2\chi_2^{RPA}\right)^2 + \left(\chi_2^{RPA} - G_1\chi_2^{RPA} - G_2\chi_1^{RPA}\right)^2} \quad (19.102)$$

where the dependence on (W, q) in the quantities appearing in the right-hand side of Equation (19.102) has been suppressed for brevity

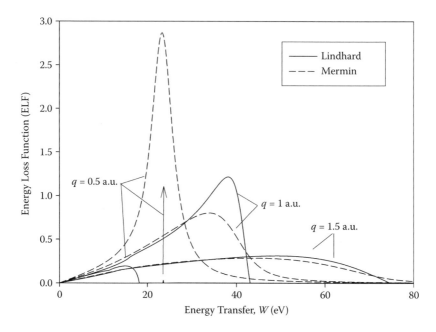

FIGURE 19.11
The energy loss function computed by the Lindhard and Mermin dielectric functions at different values of momentum transfer. A damping constant of $\gamma = 5$ eV is used in the Mermin model. Calculations are based on typical values for organic matter: $E_{pl} = 20$ eV or $r_s = 1.77$.

Since the LFC modifies the dielectric function, it would also influence the plasmon dispersion relations. Within the LFC theory, the dispersion relation for the plasmon energy up to second order in q reads:

$$E_{pl}(q) = E_{pl}^{RPA}(q) + E_{pl}^{LFC}(q) \tag{19.103}$$

with $E_{pl}^{RPA}(q)$ being the Lindhard dispersion of Equation (19.88) and $E_{pl}^{LFC}(q)$ is obtained from

$$E_{pl}^{LFC}(q) = -\frac{1}{2}E_{pl}\, q^2 \lim_{q \to 0}\left[\frac{G_1(W \approx E_{pl}, q)}{q^2}\right] \tag{19.104}$$

The corresponding relation for the plasmon line width is

$$\Delta E_{pl}(q) = \underbrace{\Delta E_{pl}^{RPA}(q)}_{=0} + \Delta E_{pl}^{LFC}(q) \tag{19.105}$$

with

$$\Delta E_{pl}^{LFC}(q) = \frac{1}{2}E_{pl}\, q^2 \lim_{q \to 0}\left[\frac{G_2(W \approx E_{pl}, q)}{q^2}\right] \tag{19.106}$$

Thus, the LFC has two important consequences for the plasmon dispersion at small q; namely, its real part, $G_1(W, q)$, causes a reduction of the plasmon energy from its RPA value according to Equations (19.103) and (19.104), while its imaginary part, $G_2(W, q)$, gives rise, via Equations (19.105) and (19.106), to plasmon damping that is neglected in RPA. Plasmon decay in the *interacting* HEG is via multipair e-h excitations, with two e-h pairs having the highest probability.

19.5.8 Static Approximation

It is natural to ask at this point what are the benefits of introducing the LFC, since it appears that what we have essentially done is to transfer all the complexities of the many-body aspects of the e-e interaction from the calculation of $\varepsilon_{HEG}(W, q)$ to the calculation of $G(W, q)$. The benefit of working with $G(W, q)$ instead of $\varepsilon_{HEG}(W, q)$ is that the former can be readily obtained under various approximations. First, we should point out that the W-dependence of the LFC, which reflects the dynamics of the e-e interaction, gives rise to a finite imaginary part, $G_2(W, q) \neq 0$, which is responsible for damping. Unfortunately, $G_2(W, q)$ is very difficult to calculate. Thus, one usually considers the LFC at the *static limit* ($W \to 0$), denoted as the static-LFC (SLFC), which is much easier to calculate and, importantly, can often be expressed *analytically*. In the static approximation we have

$$\left.\begin{array}{l} \lim_{W \to 0} G_1(W, q) \neq 0 \\[2mm] \lim_{W \to 0} G_2(W, q) = 0 \end{array}\right\} \Rightarrow \lim_{W \to 0} G(W, q) = \lim_{W \to 0} G_1(W, q) \equiv G(q) \qquad (19.107)$$

and the ELF of Equation (19.102) is simplified to

$$\mathrm{Im}\left[\frac{-1}{\varepsilon_{HEG}(W, q)}\right]_{SLFC} = \frac{\chi_2^{RPA}(W, q)}{\left\{1 + [1 - G(q)]\chi_1^{RPA}(W, q)\right\}^2 + \left\{[1 - G(q)]\chi_2^{RPA}(W, q)\right\}^2} \qquad (19.108)$$

But how useful is the static approximation in inelastic scattering where, by definition, $W \neq 0$? It turns out that, for not too large W (perhaps up to $\sim E_{pl}$), and assuming that the W-dependence of the LFC is weak, the static limit provides a reasonable first approximation to the effect of e-e interactions in HEG. In the static limit Equation (19.104) reads:

$$E_{pl}^{SLFC}(q) = -\frac{1}{2} E_{pl}\, q^2 \lim_{q \to 0}\left[\frac{G(q)}{q^2}\right] \qquad (19.109)$$

where

$$\lim_{q \to 0} G(q) = A\, (q/q_F)^2 \qquad (19.110)$$

with A being a positive dimensionless parameter that depends upon r_s (i.e., it is material dependent). Then, the plasmon dispersion relation of Equation (19.103) takes the form

$$E_{pl}(q) = E_{pl} + (\alpha_{\text{RPA}} - \alpha_{\text{SLFC}})\frac{q^2}{2m} \qquad (19.111)$$

with α_{RPA} obtained from Equation (19.89) and

$$\alpha_{\text{SLFC}} = \left(\frac{1}{2}\frac{E_{pl}}{E_{\text{F}}}\right)A \qquad (19.112)$$

With respect to the plasmon line width, it follows from Equations (19.106) and (19.107) that in the static limit $\Delta E_{pl}^{\text{SLFC}}(q) = 0$.

There exist several analytic expressions for $G(q)$. We should note in passing that interest in finding analytic expressions for the LFC (mostly for its static limit) has been intensified since the advent of TDDFT, whereby the calculation of electronic excitations crucially depends upon the availability of good approximations for the so-called exchange-correlation (XC) kernel $f_{\text{XC}}(W, q)$ of the interacting HEG, which is directly related to the (many-body) LFC through the relation $G(W,q) = -(q^2/4\pi)f_{\text{XC}}(W,q)$. The first evaluation of $G(q)$ was made by Hubbard (1958), who obtained a very simple approximation for the exchange part of the correction while neglecting the correlation contribution. The Hubbard SLFC for exchange reads:

$$G_{\text{H}}(q) = \frac{q^2}{2(q^2 + q_{\text{F}}^2)} \qquad (19.113)$$

Using Equation (19.110) we obtain $A = 1/2$ in the Hubbard approximation (independent of r_s).

Another commonly used expression for the exchange-only SLFC is that of Nozieres and Pines (1959), which leads to $A = 3/20$. A Hubbard type expression for both exchange and correlation can be obtained from the works of Kleinman (1967) and Langreth (1969). The Kleinman-Langreth SLFC for exchange-correlation reads:

$$G_{\text{KL}}(q) = \frac{1}{2}\left(\frac{q^2}{q^2 + q_{\text{F}}^2 + q_{\text{TF}}^2} + \frac{q^2}{q_{\text{F}}^2 + q_{\text{TF}}^2}\right) \qquad (19.114)$$

where q_{TF} is the Thomas-Fermi momentum defined through $q_{\text{TF}}^2 = 4q_{\text{F}}/\pi a_{\text{B}}$.

In Figure 19.12 we depict the variation of the SLFC, $G(q)$, with q within the Hubbard and the Kleinman-Langreth approximations for two materials of different electronic densities associated with electron gas parameters $r_s = 1.5$ and $r_s = 2$, respectively (which corresponds to the range of organic materials). For both models, $G(q)$ increases as q^2 at very small q

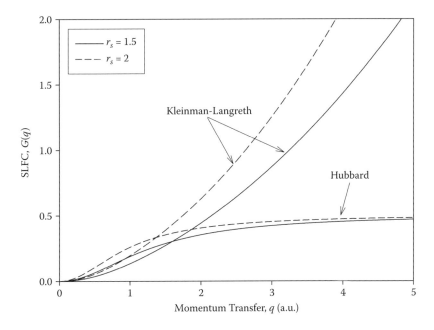

FIGURE 19.12
The Hubbard and Kleinman-Langreth static local field corrections (SLFCs) for two materials with different electron gas parameters.

in accordance to Equation (19.110). However, a very different behavior is observed with increasing q depending on whether correlation is considered. Specifically, the Hubbard exchange-only SLFC reaches asymptotically a plateau value of ½, whereas the Kleinman-Langreth expression exhibits a continous rise that is in accordance to the theoretically expected limit at high-q $G(q) \approx C (q/q_F)^2$, with the value of C being dependent on the correlation energy of the system.

In Figure 19.13 we compare the plasmon dispersion in the RPA calculated from Equation (19.88) with the plasmon dispersion as modified by the Hubbard and Kleinman-Langreth SLFCs according to Equation (19.111). It can be seen that the use of the SLFC shifts the plasmon dispersion downwards.

The effect of the Hubbard and Kleinman-Langreth SLFCs in the ELF is shown in Figure 19.14 for three different values of q. The inset depicts a situation where $q < q_c$ (i.e., Landau damping is absent) and there is a d-like plasmon peak to the right of the single-electron excitation spectrum.

There exist several more Hubbard type SLFCs in the literature as well as more refined models of increased sophistication and mathematical complexity. For an in-depth discussion of electron liquid theory one may consult the book by Giuliani and Vignale (2005).

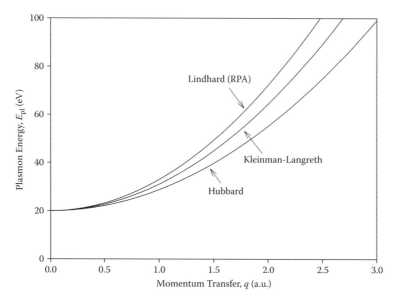

FIGURE 19.13
The plasmon dispersion calculated from the Lindhard (RPA) model and as modified by the Hubbard and Kleinman-Langreth static local field corrections (SLFCs). Calculations are based on typical values for organic matter: $E_{pl} = 20$ eV or $r_s = 1.77$.

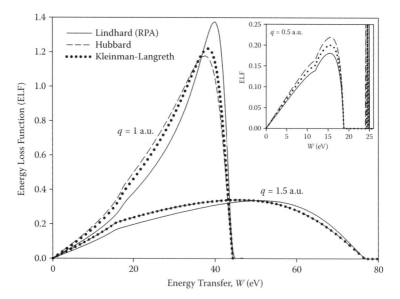

FIGURE 19.14
The effect of the Hubbard and Kleinman-Langreth static local field corrections (SLFCs) upon the Lindhard (RPA) energy loss function for different values of momentum transfer. The inset depicts a situation where the momentum transfer is below the Landau cutoff. Calculations are based on typical values for organic matter: $E_{pl} = 20$ eV or $r_s = 1.77$.

19.6 Optical Data Models

The use of experimental optical data to compute the inelastic scattering of electrons in solids was first suggested, independently, by Powell (1968, 1974) and Howie and Stern (1972). Their suggestion formed the basis of optical data models whereby the optical limit ($q \to 0$) of the dielectric function or the ELF is determined from experimental data at $q \approx 0$, while extrapolation to $q \neq 0$, where experimental data are limited (or nonexisting), is provided by theoretically motivated so-called extension algorithms (Fernandez-Varea et al. 1993). In the simplest case, an extension algorithm is a dispersion relation of energy transfer as a single-valued function of momentum transfer. Extension algorithms are commonly based on the HEG theory. Although the HEG theory applies formally to intraband (rather than interband) excitations, it can be used as an extension algorithm even for materials with a nonzero band-gap (e.g., insulators) since the value of the latter is typically well below the incident particle energy (Fernandez-Varea et al. 1993). On the other hand, the application of the HEG theory to the inner shells, which have a large excitation threshold (typically above 100 eV), is clearly problematic. Thus, in cases where the inner shells play an important role (e.g., in stopping power) the HEG theory is commonly restricted to the valence part.

Optical data models currently represent the state of the art for inelastic scattering calculations in solids at not too low particle energies (Powell and Jablonski 1999). Perhaps their main advantage is that they can be applied to any material (metal, semiconductor, or insulator) and over a wide range of incident particle energies. Moreover, the use of experimental data ensures (within experimental uncertainties) that both single-electron and collective excitations are accounted for in a material-specific manner. By virtue of the Bethe theory, optical data models are expected to be more reliable at high- than at low-incident energies since, with increasing particle velocity, the main contribution to inelastic scattering comes from collisions with $q \approx 0$ via the leading term in Bethe's asymptote. On the other hand, the quality of the extension algorithm will be increasingly important as the particle velocity decreases due to the contribution of collisions with $q \neq 0$ via higher-order terms (in $1/T$) in the Bethe asymptote.

19.6.1 Optical Limit

Experimental data at the optical limit ($q \approx 0$; denoted as optical data) for condensed materials are commonly available up to ~50–100 eV (i.e., over the valence excitation range) from either optical measurements or electron energy loss spectroscopy (EELS). In the former case one measures the index of refraction, $n = n(W)$, and extinction coefficient, $\kappa = \kappa(W)$, of the material from which the dielectric function at $q \approx 0$ can be determined through the relations

$$\varepsilon(W,q \approx 0)_{\text{val}} = (n + i\kappa)^2 \Rightarrow \begin{cases} \varepsilon_1(W,0) = n^2 - \kappa^2 \\ \varepsilon_2(W,0) = 2n\kappa \end{cases} \qquad (19.115)$$

where the subscript val denotes the valence (or outer) electrons of the material. It is then straightforward to compute the ELF (also at the optical limit) by

$$\text{Im}\left[\frac{-1}{\varepsilon(W,q \approx 0)}\right]_{\text{val}} = \text{Im}\left[\frac{-1}{(n + i\kappa)^2}\right] = \frac{2n\kappa}{(n^2 - \kappa^2)^2 + (2n\kappa)^2} \qquad (19.116)$$

EXAMPLE 19.6

Calculate the ELF (at $q = 0$), the OOS (per electron), and the photoabsorption cross section σ_{pk} (per electron) of liquid water at energy transfer of $W = 21$ eV using the values for the optical constants from Table I of Heller et al. (1974).

SOLUTION 19.6

At the energy transfer of 21 eV we are at the region of valence electron excitations, so we can use Equation (19.116) for calculating the ELF. Substituting the values of the optical constants (at 21 eV) $n = 0.77$ and $\kappa = 0.39$ in Equation (19.116) we obtain

$$\text{Im}\left[\frac{-1}{\varepsilon(W,0)}\right] = \frac{2 \times 0.77 \times 0.39}{(0.77^2 - 0.39^2)^2 + (2 \times 0.77 \times 0.39)^2} = 1.08$$

Then, Equation (19.36) can be solved for the GOS (at $Q = q^2/2m = 0$) to obtain (using $E_{pl} = 21.4$ eV from Example 19.2)

$$\frac{df(W,0)}{dW} = \frac{2}{\pi} \frac{W}{E_{pl}^2} \text{Im}\left[\frac{-1}{\varepsilon(W,0)}\right]$$

$$= \frac{2}{3.14} \frac{21\,\text{eV}}{(21.4\,\text{eV})^2} 1.08 = 0.0315\,\text{eV}^{-1} \text{ per electron}$$

Substituting to Equation (19.24) we obtain for the photoabsorption cross section of liquid water at 21 eV:

$$\sigma_{ph} = 4 \times 3.14^2 \times \frac{1}{137} \times (0.529 \times 10^{-8}\,\text{cm})^2 \times 13.6\,\text{eV} \times 0.0315\,\text{eV}^{-1}$$

$$= 3.45 \times 10^{-18}\,\text{cm}^2 \text{ per electron}$$

Alternatively, information about the ELF at $q \approx 0$ can be obtained by EELS measurements at nearly forward scattering, i.e., at vanishing q. This is accomplished by electron beam irradiation of the sample using a transmission electron microscope (TEM) and collecting only those electrons scattered within the angular range $0 < \theta \le \theta_c$, where the cutoff scattering angle θ_c can be determined by the experimental setup to be of only a few degrees. Under this condition it is reasonable to assume that the EELS data represent the *energy loss spectrum* of the incident electrons (of fixed energy) in a *single* inelastic collision within the target material, i.e., the differential-in-energy-loss inelastic scattering cross section. Then, under the conditions $\theta_c \ll 1$ and $W \ll T$ the following relation holds:

$$\mathrm{Im}\left[\frac{-1}{\varepsilon(W, q \approx 0)}\right]_{\mathrm{val}} \approx \frac{\pi a_B T}{\ln[1 + (2T\theta_c/W)^2]} \times \frac{d\Lambda_{\mathrm{EELS}}}{dW}\bigg|_{\theta \le \theta_c} \qquad (19.117)$$

Although $\mathrm{Im}[-1/\varepsilon(W, q \approx 0)]$ has its maximum at the valence excitation range where condensed phase effects are dominant, for high incident energies (e.g., keV electrons or MeV protons) the valence excitations represent only a small part of the kinematically accessible excitation range. To extend $\mathrm{Im}[-1/\varepsilon(W, q \approx 0)]$ to core electron excitations that fall in the x-ray range (i.e., above ~100 eV) it is generally assumed that at such high-energy transfers screening effects are negligible; i.e., the excitations in the condensed phase are similar to those of the gas phase. Then, from Equations (19.30) and (19.35) and assuming the absence of screening, i.e., $|\varepsilon(W, q)| \approx 1$, we obtain for the core excitations:

$$\mathrm{Im}[-1/\varepsilon(W)]_{\mathrm{core}} \approx \mathrm{Im}\,\varepsilon(W)_{\mathrm{core}} = \left(2\pi^2 e^4 a_B N\right) W^{-1}\, df_x(W)/dW \qquad (19.118)$$

with $df_x(W)/dW$ being the OOS *per atom* in the x-ray range. Using now the relation given by Equation (19.24) between the photoabsorption cross section and the atomic OOS we obtain the more compact expression:

$$\mathrm{Im}\left[\frac{-1}{\varepsilon(W, q \approx 0)}\right]_{\mathrm{core}} \approx \varepsilon_2(W, 0)_{\mathrm{core}} = c\hbar W^{-1}\mu_x \qquad (19.119)$$

with $\mu_x = N\sigma_x$ being the linear absorption coefficient in the x-ray range and σ_x being the corresponding atomic photoabsorption cross section (recall that N is the atomic density). The values of μ_x (or σ_x) can be determined either empirically or, as is often the case, via theoretical atomic calculations.

EXAMPLE 19.7

Calculate the ELF (at $q = 0$), the OOS (per electron), and the photoabsorption cross section (per electron) of liquid water at the energy transfer of $W = 1$ keV using the photoelectric coefficients of the XCOM database of NIST.

SOLUTION 19.7

From the XCOM database, the mass photoabsorption coefficient of water at 1 keV is $\mu/\rho = 4.076 \times 10^3$ cm^{2}g^{-1}. Thus, for liquid water of mass density $\rho = 1$ g cm^{-3}, the linear absorption coefficient of Equation (19.119) is $\mu_x = 4.076 \times 10^3$ cm^{-1}. Since the energy transfer of 1 keV is above the K-edge (~540 eV) of water, we are at the region of core electron excitations where screening effects can be neglected. The ELF should then be calculated from Equation (19.119):

$$\text{Im}\left[\frac{-1}{\varepsilon(W,0)}\right]_{W=1\,\text{keV}} = 3 \times 10^{10}\,\text{cm s}^{-1} \times 6.58 \times 10^{-16}\,\text{eV s}\,\frac{4.076 \times 10^3\,\text{cm}^{-1}}{1000\,\text{eV}}$$

$$= 8.05 \times 10^{-5}$$

Substituting to Equation (19.118) and solving for the OOS after noting that $2\pi^2 e^4 a_B\, N\, Z = \pi E_{pl}^2/2$ (multiplication by Z is necessary for expressing the OOS per electron) we obtain

$$\frac{df(W,0)}{dW} = \frac{2}{\pi}\frac{W}{E_{pl}^2}\,\text{Im}\left[\frac{-1}{\varepsilon(W,0)}\right]$$

$$= \frac{2}{3.14}\frac{1000\,\text{eV}}{(21\,\text{eV})^2}(8.05 \times 10^{-5}) = 1.16 \times 10^{-4}\,\text{eV}^{-1}\,\text{per electron}$$

For the photoabsorption cross section (per electron) we can use Equation (19.24) or the relation $\mu_x = N_e\,\sigma_x$. From the latter expression and noting that (Example 19.1) $N_e = 3.34 \times 10^{23}$ cm^{-3} for liquid water we obtain

$$\sigma_x = \frac{\mu_x}{N_e} = \frac{4.076 \times 10^3\,\text{cm}^{-1}}{3.34 \times 10^{23}\,\text{cm}^{-3}} = 1.22 \times 10^{-20}\,\text{cm}^2\,\text{per electron}$$

Although the use of *atomic* coefficients is strictly correct for the gas phase, the justification that is commonly put forward for extending their application to the condensed phase is that for values of W that fall in the x-ray range, the energy transfer is much larger than the plasmon energy, and therefore, to a good approximation, the excitation spectrum is insensitive to whether the system is in the gas or condensed phase. In fact, for $W \gg E_{pl}$ the condition $|\varepsilon(W)| \approx 1$ holds. Thus, the optical data model (ODM) ELF at $q = 0$ is obtained from the approximation

$$\text{Im}\left[\frac{-1}{\varepsilon(W,q \approx 0)}\right]_{\text{ODM}} \approx \text{Im}\left[\frac{-1}{\varepsilon(W,0)}\right]_{\text{val-solid}} + \text{Im}\,\varepsilon(W,0)_{\text{core-atomic}} \quad (19.120)$$

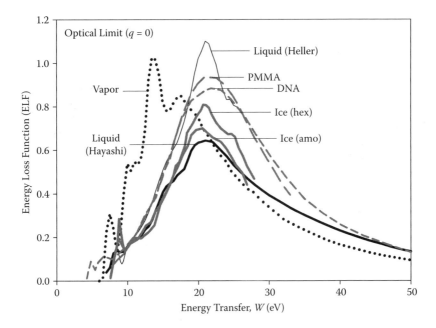

FIGURE 19.15
The energy loss function for several biomaterials at the optical limit ($q \approx 0$) obtained from experimental optical data. Liquid water data from Heller et al. (1974) and Hayashi et al. (2000), dry DNA data from Inagaki et al. (1974), PMMA data from Ritsko et al. (1978), water ice data from Kobayashi (1983), and water vapor data from Chan et al. (1993).

It is important to recognize that the above procedure is not restricted to the type of material (metal, semiconductor, or insulator) but only to the availability of the relevant optical data.

In Figure 19.15 we show the ELF obtained from optical data ($q \approx 0$) for several biomaterials. Contrary to water vapour, which exhibits several sharp peaks corresponding to discrete excitations and the ionization (binding) energies of the molecular orbitals, the spectra of the condensed materials have a characteristic broad maximum at about 21 eV, which coincides well with their nominal plasmon energy calculated from Equation (19.32), and some minor structure in the range 5–15 eV.

It is clear that any ODM is based on the combination of information from different sources as well as interpolation or extrapolation across the excitation spectrum. A useful tool for checking the internal consistency of an ODM is offered by the sum rules. The most commonly used is the *f*-sum rule of Equation (19.37), which in the present case reads:

$$\frac{2}{\pi E_{pl}^2} \int_0^\infty W \operatorname{Im}\left[\frac{-1}{\varepsilon(W, q \approx 0)}\right]_{ODM} dW = 1 \qquad (19.121)$$

For most biomaterials that consist of low-Z elements, the value of the integral of Equation (19.121) doesn't practically change when the upper limit is set beyond several hundred keV. Due to the multiplication of the ELF by W, the f-sum rule of Equation (19.121) is mostly sensitive to the quality of the ODM in the region of large W, practically above ~50–100 eV, i.e., to the x-ray data that correspond to core excitations.

However, the whole idea of invoking the dielectric formalism was to account for condensed phase effects that are pronounced in the range of valence excitations, i.e., below ~50–100 eV. Thus, the f-sum rule does not provide a very stringent test for the internal consistency of the ODM over the important range of valence excitations.

Another commonly encountered consistency check of an ODM pertains to the calculation of the I value of Bethe's stopping power from Equation (19.54) and its comparison with experimentally determined I values. From inspection of Equation (19.54) it follows that the calculation of the I value would sensitively depend upon the contribution of high-W excitations (e.g., ionizations of inner shell electrons). For low-Z materials, the I value reaches its asymptotical value in the 10–100 keV energy loss range—that is, at excitation energies much larger than the K-shell binding energy of these materials. This has two important implications. First, the I value test (similar to the f-sum rule test) may hide possible inaccuracies of the ODM over the energy range of valence excitations. Second, discrepancies between different ODMs in the region of core electron excitations (e.g., of the oxygen K-shell in the case of water) may result in sizable differences in the calculated I value. Thus, recent theoretical estimates of the I value of liquid water (as reviewed by Emfietzoglou et al. 2009) based on ODMs differ by nearly 10 eV, ranging from 72.5 to 82.4 eV. Such differences in the I value are very important in various contexts, such as, for example, in the dosimetry of ion beams and, more specifically, in the calculation of the ion's depth-dose profile (and its Bragg peak) in the medium via Bethe's theory (Equation 19.56).

A sum rule sensitive to the low part of the excitation spectrum is provided by the KK relations. Taking the static ($W \to 0$) and optical ($q \to 0$) limit of Equation (19.39a) we obtain

$$\frac{2}{\pi} \int_0^\infty \mathrm{Im} \left[\frac{-1}{\varepsilon(W, q \approx 0)} \right]_{\mathrm{ODM}} \frac{dW}{W} + \mathrm{Re} \left[\frac{1}{\varepsilon(W=0, q=0)} \right] = 1 \qquad (19.122)$$

which is often called the KK-sum rule (Tanuma et al. 1993a). Note that in Equation (19.122) the ELF is multiplied by W^{-1}, which makes the integral most sensitive to the behavior of ELF at small values of W. For conductors $\mathrm{Re}[1/\varepsilon(0,0)] \equiv 1/\varepsilon_1(0,0) = 0$ and Equation (19.122) is then the so-called perfect screening sum rule. For semiconductors or insulators, according to Equation (19.115) and the relation $n(0) \gg \kappa(0)$, we have $\mathrm{Re}[1/\varepsilon(0,0)] \approx 1/n^2(0)$.

19.6.2 Models Based on the Drude Dielectric Function

The extended Drude model was first suggested in the mid-1970s by Ritchie and coworkers (Hamm et al. 1975; Ritchie and Howie 1977) for modeling in a consistent manner the ELF of condensed materials over the complete range of energy and momentum transfer, i.e., the Bethe surface. There are two different versions of the extended Drude model. In the more elaborate version one starts from the optical data for the dielectric function $\varepsilon(W, q \approx 0)$, whereas in the simpler version one works directly with the optical data for the ELF, $\text{Im}[-1/\varepsilon(W, q \approx 0)]$. The former version has been the basis of the Oak Ridge Electron Code (OREC) Monte Carlo code for electron transport in liquid water (Hamm et al. 1975; Ritchie et al. 1978, 1991), whereas the latter version (Ritchie and Howie 1977), often referred to as the Ritchie-Howie scheme, has been used extensively by theorists in materials science (e.g., Yubero et al. 1996). As will be shown below, common to both models is the analytic parameterization of optical data by Drude-like functions.

19.6.2.1 OREC Version

Due to the direct connection of the imaginary part of the dielectric function with OOS via Equation (19.30), it is assumed that the experimental optical data for $\varepsilon_2(W, q \approx 0)$ can be partitioned to the principal absorption modes of the system. Thus, a superposition of modified Drude (MD) dielectric functions of the form of Equation (19.76) is used to parameterize the optical data as follows:

$$\varepsilon_2^{\exp}(W, q \approx 0) \approx \sum_n \text{Im}\, \varepsilon_{MD}^{(n)}(W, 0) = E_{pl}^2 \sum_n D(W; f_n, E_n, \gamma_n) \quad (19.123)$$

with

$$D(W; f_n, E_n, \gamma_n) \equiv \frac{f_n\, W\, \gamma_n}{(E_n^2 - W^2)^2 + (W\, \gamma_n)^2} \quad (19.124)$$

being the so-called *Drude function*.

From the KK relation, Equation (19.38a), the experimental optical data for $\varepsilon_1(W, q = 0)$ can also be represented analytically as follows:

$$\varepsilon_1^{\exp}(W, q \approx 0) \approx 1 + E_{pl}^2 \sum_n D_{KK}(W; f_n, E_n, \gamma_n) \quad (19.125)$$

with

$$D_{KK}(W; f_n, E_n, \gamma_n) = \frac{f_n(E_n^2 - W^2)}{(E_n^2 - W^2)^2 + (W\, \gamma_n)^2} \quad (19.126)$$

being the KK pair of Equation (19.124). Then, using Equations (19.124) and (19.126), we also get:

$$
\mathrm{Im}\left[\frac{-1}{\varepsilon(W,q\approx0)}\right]_{\mathrm{exp}} \approx \frac{E_{pl}^2 \sum_n D(W;f_n,E_n,\gamma_n)}{[1+E_{pl}^2 \sum_n D_{\mathrm{KK}}(W;f_n,E_n,\gamma_n)]^2 + [E_{pl}^2 \sum_n D(W;f_n,E_n,\gamma_n)]^2}
$$

(19.127)

The Drude coefficients f_n, E_n, and γ_n can be formally interpreted as the optical oscillator strength, resonance energy, and damping constant of the nth electronic transition of the target. The values of the Drude coefficients are found empirically by fitting the experimental optical data via Equations (19.123), (19.125), and (19.127) under the constraints of the f- and KK-sum rules, Equations (19.121) and (19.222), respectively.

To improve upon the asymptotic properties of the Drude function of Equation (19.124) it was later suggested (Ritchie et al. 1978, 1991) to replace it by the so-called *derivative* Drude function (denoted here by $\tilde{D}$), which has the following analytic form (at $q = 0$):

$$
\tilde{D}(W;f_n,E_n,\gamma_n) \equiv \frac{2f_n(W\gamma_n)^3}{\left[\left(E_n^2-W^2\right)^2 + (W\gamma_n)^2\right]^2}
$$

(19.128)

Apparently, the *derivative* Drude function ($\tilde{D}$) is not exactly obtained by differentiation from the ordinary Drude functions (D), but it shares with the latter important analytic properties, such as, for example, the value of the f-sum rule integral as well as having an analytic KK pair. The latter is used to model ε_1 and $\mathrm{Im}(-1/\varepsilon)$ by replacing D_{KK} with

$$
\tilde{D}_{\mathrm{KK}}(W;f_n,E_n,\gamma_n) = \frac{f_n\left(E_n^2-W^2\right)\left[\left(E_n^2-W^2\right)^2 + 3W^2\gamma_n^2\right]}{\left[\left(E_n^2-W^2\right)^2 + (W\gamma_n)^2\right]^2}
$$

(19.129)

The main improvement brought about by the use of the derivative Drude function concerns the asymptotic properties of the dielectric function at large excitation energies W. Specifically, it can be seen that $D \propto W^{-3}$, whereas $\tilde{D} \propto W^{-5}$. However, at large W we expect $\mathrm{Im}[-1/\varepsilon(W)] \approx \mathrm{Im}\,\varepsilon(W)$ so we can use as a guide the theoretically expected asymptotic trend predicted by the atomic OOS, which is $df(W)/dW \propto W^{-3.5}$, so $\varepsilon_2(W) \propto W^{-1}df(W)/dW \propto W^{-4.5}$. Thus, the derivative Drude function is better conditioned to expected asymptotic trends than the ordinary Drude function.

As it has been pointed out by Dingfelder et al. (1998, 1999, 2000), the dielectric function obtained from Equation (19.123) takes finite values for any nonvanishing energy transfer ($W \neq 0$). This is clearly unrealistic since the minimum excitation energy should be at least equal to the band-gap energy (e.g., ~7–8 eV for liquid water). The effect is more pronounced for ionizations, where the energy transfer should exceed the binding energy of the corresponding shell. To solve this problem Dingfelder and coworkers suggested to represent the ionizations by "truncated" Drude functions (denoted here as D^*) of the form

$$D^*(W; f_n, E_n, \gamma_n, \Delta_n) = \int_{E_n-\Delta_n}^{E_n+\Delta_n} D(W; f_n, \Omega, \gamma_n) G(\Omega, E_n, \Delta_n) \Theta(\Omega - W) d\Omega \qquad (19.130)$$

where $\Theta()$ denotes the step function and

$$G(\Omega, E_n, \Delta_n) = \exp\left\{ -\frac{(\Omega - E_n)^2}{2\Delta_n^2} \right\} \qquad (19.131)$$

is a Gaussian "smearing" function at the ionization thresholds E_n. The value of Δ_n, which represents the width of the nth band, was determined empirically to be $\Delta_n \approx 1$ eV (the same value is assumed for all n). In addition, Dingfelder and coworkers suggested the representation of the discrete excitations by a derivative Drude function that is more sharply peaked and therefore better suited for the sharp excitation peaks that often appear at low-energy transfers. Then, in place of Equation (19.123) we obtain

$$\varepsilon_2^{exp}(W, q \approx 0) \approx E_{pl}^2 \left[\overset{excit}{\underset{n}{\sum}} \tilde{D}(W; f_n, E_n, \gamma_n) + \overset{ioniz}{\underset{n}{\sum}} D^*(W; f_n, E_n, \gamma_n, \Delta_n) \right] \qquad (19.132)$$

$\varepsilon_1(W, q = 0)$ can be determined from the KK relation; however, due to the Ω-integration, the KK pair of the truncated Drude function of Equation (19.130) cannot be expressed in an analytic form, but it has to be calculated numerically.

In Figure 19.16 we present a Drude parameterization (Emfietzoglou and Nikjoo 2005) of the experimental optical data ($q \approx 0$) of liquid water obtained via inelastic x-ray scattering (IXS) spectroscopy (Hayashi et al. 2000) using the OREC prescription.

19.6.2.2 Ritchie-Howie Version

A simpler scheme is suggested by Ritchie and Howie (1977) based on a Drude parameterization of the experimental optical data for the ELF using the expression

$$\text{Im}\left[\frac{-1}{\varepsilon(W, q \approx 0)} \right]_{exp} \approx \sum_n \frac{A_n}{E_n^2} \text{Im}\left[\frac{-1}{\varepsilon_D(W; E_n, \gamma_n)} \right] \qquad (19.133)$$

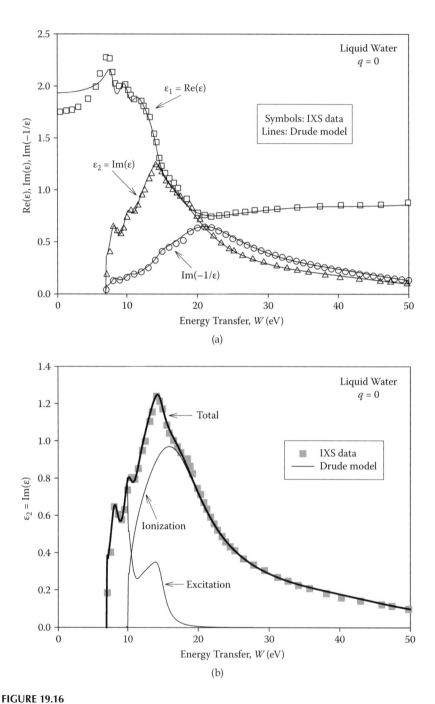

FIGURE 19.16
Drude parameterization of the experimental IXS optical data ($q \approx 0$) of liquid water (a) and the contribution of excitations and ionizations to Im(ε) (b).

where according to Equation (19.70)

$$\text{Im}\left[\frac{-1}{\varepsilon_D(W;E_n,\gamma_n)}\right] = \frac{E_n^2\,\gamma_n W}{\left(E_n^2 - W^2\right)^2 + (W\gamma_n)^2} \qquad (19.134)$$

Note that if we set $A_n = f_n E_{pl}^2$, then Equation (19.133) can be recast as

$$\text{Im}\left[\frac{-1}{\varepsilon(W,q\approx 0)}\right]_{\text{exp}} \approx E_{pl}^2 \sum_n D(W;f_n,E_n,\gamma_n) \qquad (19.135)$$

where the right-hand side is identical to that of Equation (19.123). The empirically determined Drude coefficients A_n, E_n, and γ_n can now be loosely interpreted as the intensity, effective energy, and damping of a "bound" plasmon associated with the nth subband. Similar to the OREC model, the values of the coefficients are obtained empirically through Equation (19.135) under the constraint of the f- and KK-sum rules for the ELF. Unlike the OREC version, however, threshold effects can be easily included in the Ritchie-Howie expression, Equation (19.133), by multiplying with $\Theta(W - E_{n,\text{cut}})$, where the cutoff energies $E_{n,\text{cut}}$ would equal the band-gap energy when n represents valence excitations in semiconductors and insulators, or the binding energy when n represents inner shell excitations.

19.6.2.3 Extension to Arbitrary q

In the extended Drude models the extension to $q \neq 0$ is made through appropriate dispersion relations for the Drude coefficients. The q-dependence of the Drude coefficients is commonly obtained via analytic dispersion relations, which are meant to account for global features of the Bethe surface. Then, for example, in the Ritchie-Howie version we obtain for arbitrary q:

$$\text{Im}\left[\frac{-1}{\varepsilon(W,q)}\right]_{\text{Ritchie-Howie}} = \sum_n \frac{A_n}{E_n^2} \text{Im}\left[\frac{-1}{\varepsilon_D\left(W;E_n(q),\gamma_n(q)\right)}\right] \qquad (19.136)$$

Note that the extended Drude dielectric function, ε_D, in Equation (19.136) is essentially the plasmon pole, ε_{PP}, dielectric function of Equation (19.96). As expected, at $q = 0$, Equation (19.136) reduces to Equation (19.133). From the properties of the Drude dielectric function, it follows that the f-sum rule in the extended Drude models will be fulfilled for any q, *independent of the form of the dispersion relations $E_n(q)$ and $\gamma_n(q)$, as long as it is fulfilled at $q = 0$.* However, in case threshold effects are introduced in Equation (19.136) via a step function, then the cutoff energies must also be made q-dependent, $E_{n,\text{cut}}(q)$, for the f-sum rule to be fulfilled at $q \neq 0$.

The simplest (and most common) suggestion is to use an RPA-like dispersion relation in Equation (19.136) by writing

$$E_n(q) = E_n + \alpha_{RPA}\frac{q^2}{2m} \tag{19.137}$$

where for biological materials $\alpha_{RPA} \approx 1$ to within 10%. Then, Equation (19.137) is correct at the two limits: at $q \to 0$, where $E_n(q) = E_n$, and also at $q \to \infty$, where $E_n(q) \approx q^2/2m$. In between the two limits Equation (19.137) provides a reasonable approximation to the RPA plasmon dispersion relation. Momentum broadening effects in the ELF are commonly neglected by assuming a nondispersive damping constant $\gamma_n(q) = \gamma_n(q = 0) \equiv \gamma_n$. On the other hand, in cases where experimental data at $q \neq 0$ are available, one can treat the quadratic dispersion coefficient in Equation (19.137) as an additional adjustable (empirical) parameter and also employ a similar dispersion relation for the damping coefficient (Ding and Shimizu 1989; Kuhr and Fitting 1999). From an analysis of the experimental IXS data at $q \neq 0$ for liquid water (Watanabe et al. 1997, 2000), Emfietzoglou et al. (2005) have proposed improved dispersion relations that account for momentum broadening and shifting effects in the ELF, in accordance with the predictions of LFC theory. The Emfietzoglou-Cucinotta-Nikjoo (ECN) corrected-RPA relations read:

$$E_n(q) = E_n + g(q)\frac{q^2}{2m} \tag{19.138}$$

where

$$g(q) = 1 - \exp(-c\,q^d) \tag{19.139}$$

and

$$\gamma_n(q) = \gamma_n + a\,q + b\,q^2 \tag{19.140}$$

with a, b, c, d being empirical constants (Emfietzoglou et al. 2005, 2008). The effect of $g(q)$ is to reduce $E_n(q)$ from its quadratic value at not too large q, thus shifting the ELF to lower excitation energies. Note that the form of Equation (19.139) leads to the correct limiting behavior at both $q \to 0$ and $q \to \infty$, where $g(q) \approx 0$ and $g(q) \approx 1$, respectively. The latter condition ensures that single-electron excitations are still properly accounted for at large q by the quadratic energy-momentum dispersion relation characteristic of the Bethe ridge. The linear-quadratic dispersion relation for $\gamma_n(q)$ induces a momentum broadening effect in the ELF at all q (even for $q < q_c$) as predicted experimentally.

The Bethe surface of liquid water computed by the extended Drude model using the RPA dispersion relation of Equation (19.137) and the ECN dispersion relations of Equations (19.138) to (19.140) is shown in Figure 19.17. These results should be compared with the experimental Bethe surface of liquid

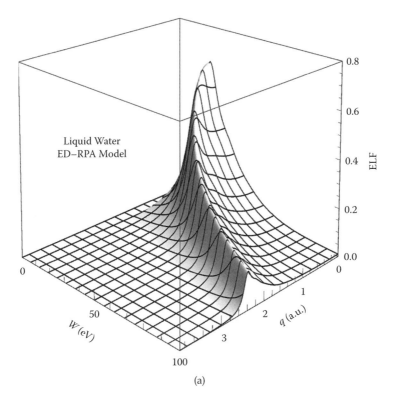

(a)

FIGURE 19.17
The Bethe surface of liquid water computed by the extended-Drude model using the RPA dispersion relation (a) and the Emfietzoglou-Cucinotta-Nikjoo improved dispersion relations (b).

water depicted in Figure 19.18. It is evident that the use of Equation (19.137) alone yields a Bethe ridge whose width is too small (although nonzero) and whose height is too large.

In Figure 19.19 the experimental ELF at a single q value is compared with the extended Drude models of Figure 19.17. The ECN dispersion relations, Equations (19.138) to (19.140), correct the standard RPA dispersion and improve the agreement with the experimental data by introducing momentum broadening and peak shifting in the ELF.

19.6.3 Models Based on the Lindhard Dielectric Function

19.6.3.1 Penn Model

Penn (1987) suggested an optical data model whereby extension of the optical data to $q \neq 0$ is made via the Lindhard dielectric function. The Penn expression for the ELF follows from a generalization of the Ritchie-Howie ELF of Equation (19.136), whereby the summation over a finite number of Drude-ELFs

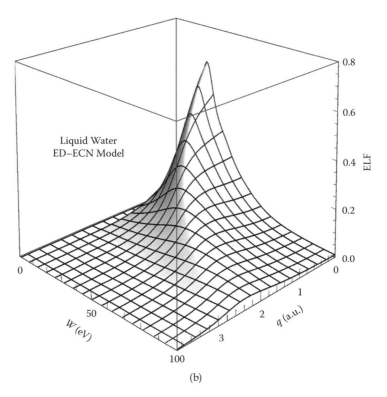

FIGURE 19.17
(Continued)

is replaced by an integration (i.e., an infinite sum) over a model ELF according to the relation

$$\text{Im}\left[\frac{-1}{\varepsilon(W,q)}\right] = \int_0^\infty A(W')\,\text{Im}\left[\frac{-1}{\varepsilon(W,q;W')}\right]_{\text{model}} dW' \qquad (19.141)$$

where the weighting factors A_n and effective plasmon energies E_n of Equation (19.136) have now been replaced by a distribution function $A(W')$ of the continuous variable W' of the excitation energy. Then, in Equation (19.141) one sets

$$\text{Im}\left[\frac{-1}{\varepsilon(W,q;W')}\right]_{\text{model}} = \text{Im}\left[\frac{-1}{\varepsilon(W,q;E_{pl})}\right]_{\text{Lindhard}} \qquad (19.142)$$

leading to

$$\text{Im}\left[\frac{-1}{\varepsilon(W,q)}\right] = \int_0^\infty A(W')\,\text{Im}\left[\frac{-1}{\varepsilon(W,q;W')}\right]_{\text{Lindhard}} dW' \qquad (19.143)$$

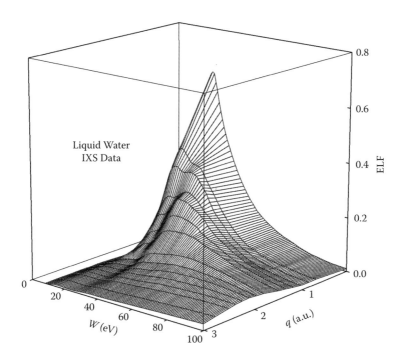

FIGURE 19.18
The Bethe surface of liquid water measured experimentally using inelastic x-ray scattering (IXS) spectroscopy by Watanabe and coworkers.

Using now the limiting form of the Lindhard-ELF,

$$\lim_{q \to 0} \mathrm{Im}\left[\frac{-1}{\varepsilon_L(W,q;E_{pl})} \right] = \frac{\pi}{2} E_{pl}\delta(W - E_{pl}) \tag{19.144}$$

in Equation (19.143) we obtain

$$\mathrm{Im}\left[\frac{-1}{\varepsilon(W,q=0)} \right] = \frac{\pi}{2} \int_0^\infty A(W')W'\delta(W - W')dW' \tag{19.145}$$

which, after solving for $A(W)$, gives

$$A(W) = \frac{2}{\pi W} \mathrm{Im}\left[\frac{-1}{\varepsilon(W,q=0)} \right] \tag{19.146}$$

Contact with experimental optical data can therefore be made by the condition

$$A(W) \approx \frac{2}{\pi W} \mathrm{Im}\left[\frac{-1}{\varepsilon(W,q \approx 0)} \right]\Bigg|_{exp} \tag{19.147}$$

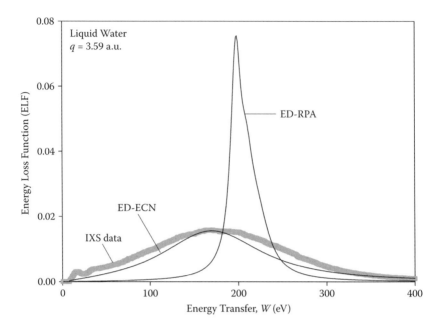

FIGURE 19.19
The experimental energy loss function of liquid water at the highest transferred momentum measured is compared with the extended-Drude model using different dispersion relations, as in Figure 19.17.

Inserting Equation (19.147) back into Equation (19.143) yields the full Penn expression for the ELF:

$$\text{Im}\left[\frac{-1}{\varepsilon(W,q)}\right]_{\text{Penn}} = \frac{2}{\pi}\int_0^\infty \text{Im}\left[\frac{-1}{\varepsilon(W',q\approx 0)}\right]_{\text{exp}} \text{Im}\left[\frac{-1}{\varepsilon(W,q;W')}\right]_{\text{Lindhard}} \frac{dW'}{W'} \quad (19.148)$$

Thus, the extension of the optical data ELF to $q \neq 0$ in the Penn model is made via the Lindhard-ELF, which results in considerable numerical work; e.g., Equation (19.148) implies an additional integration over W' for calculating inelastic scattering magnitudes. The use of the Lindhard dielectric function of the HEG assumes zero excitation threshold, thus allowing energy losses even below the band-gap energy. This property of the Penn model can be particularly problematic when applied to inner shells with large binding energies. Thus, the Penn model is not recommended for core excitations. The main advantage of the Penn model is that the ELF is automatically extended to $q \neq 0$ through the analytic properties of the Lindhard dielectric function without the need of further approximations about dispersion. The Penn model forms the basis of the widely used Tanuma-Powell-Penn (TPP) predictive formula for the electron inelastic mean free path in solids (see Tanuma et al. 1993b, 2003).

19.6.3.2 Ashley Model

Ashley (1988, 1990) proposed an approximation to the Penn model whereby the Lindhard dielectric function is replaced by the one-mode approximation, i.e., the zero-damping limit of the plasmon pole approximation. This change significantly simplifies the calculations given that the integral form of Equation (19.141) makes it particularly convenient to approximate $\mathrm{Im}[-1/\varepsilon(W,q;W')]_{\mathrm{model}}$ by a delta-function. So, according to Ashley the following relation is used:

$$\mathrm{Im}\left[\frac{-1}{\varepsilon(W,q;W')}\right]_{\mathrm{model}} = \lim_{\gamma(q)\to 0}\mathrm{Im}\left[\frac{-1}{\varepsilon_{PP}(W;E_{pl}(q),\gamma(q))}\right] = \frac{\pi}{2}\frac{E_{pl}^2}{E_{pl}(q)}\delta\left(W-E_{pl}(q)\right)$$

(19.149)

Inserting Equation (19.149) into Equation (19.141) we obtain

$$\mathrm{Im}\left[\frac{-1}{\varepsilon(W,q)}\right] = \frac{\pi}{2}\int_0^\infty A(W')\frac{W'^2}{W}\delta\left(W-W'(q)\right)dW'$$

(19.150)

where $A(W')$ is connected with the optical data by means of Equation (19.147), as it can be trivially verified by evaluating Equation (19.150) at the limit $q = 0$ using also $W'(q = 0) \equiv W'$.

Inserting Equation (19.147) into Equation (19.150) yields

$$\mathrm{Im}\left[\frac{-1}{\varepsilon(W,q)}\right] = \int_0^\infty \mathrm{Im}\left[\frac{-1}{\varepsilon(W',q\approx 0)}\right]_{\mathrm{exp}}\frac{W'}{W}\delta(W-W'(q))dW'$$

(19.151)

Thus, the extension of the optical data ELF to $q \neq 0$ in the Ashley model is essentially made via the d-function $\delta(W - W'(q))$. From the properties of the delta-function, Equation (19.151) can be recast in the following compact form:

$$\mathrm{Im}\left[\frac{-1}{\varepsilon(W,q)}\right] = \frac{W_0}{W}\mathrm{Im}\left[\frac{-1}{\varepsilon(W_0,q\approx 0)}\right]_{\mathrm{exp}}$$

(19.152)

where W_0 is the positive solution of the dispersion relation:

$$E(q;W_0) = W$$

(19.153)

To simplify the numerical work, Ashley used the Lindhard dispersion relation with $\alpha_{RPA} = 1$, which, after replacing E_{pl} by W_0, leads to the solution

$$W_0 = W - Q$$

(19.154)

Inserting Equation (19.154) into Equation (19.152) we obtain Ashley's final expression for the ELF:

$$\text{Im}\left[\frac{-1}{\varepsilon(W,q)}\right]_{\text{Ashley}} = \frac{W-Q}{W}\,\text{Im}\left[\frac{-1}{\varepsilon(W-Q,q\approx 0)}\right]_{\text{exp}}\Theta(W-Q) \quad (19.155)$$

The main advantage of the Ashley model is its computational simplicity. For example, the integration of the ELF over q can now be done analytically. On the other hand, a drawback of the Ashley model is that it shifts the excitation threshold to twice its normal value, e.g., twice the band-gap energy or twice the inner shell binding energy. This is perhaps not important for valence excitations where the band-gap energy is typically smaller than 10 eV, but it can introduce a sizable error in calculations for inner shells that have binding energies that are often comparable to the incident electron energy.

19.6.4 Models Based on the Mermin Dielectric Function

Abril et al. (1998, 2010) have proposed an extension algorithm, denoted as MELF-GOS, whereby valence excitations are dispersed according to the analytic properties of the Mermin ELF (MELF) and core excitations according to the hydrogenic GOS function. The MELF description is essentially obtained from the Ritchie-Howie scheme by replacing the Drude dielectric function with the Mermin dielectric function, leading to the following expression for the ELF:

$$\text{Im}\left[\frac{-1}{\varepsilon(W,q)}\right]_{\text{MELF}} = \sum_n \frac{A_n}{E_n^2}\,\text{Im}\left[\frac{-1}{\varepsilon_{\text{M}}(W,q;E_n,\gamma_n)}\right] \quad (19.156)$$

where $\varepsilon_{\text{M}}(W,q;E_n,\gamma_n)$ is the Mermin dielectric function of Equation (19.93). The values of the model parameters A_n, E_n, and γ_n in Equation (19.156) are determined by the fit to experimental optical data using the condition

$$\text{Im}\left[\frac{-1}{\varepsilon(W,q\approx 0)}\right]_{\text{exp}} \approx \text{Im}\left[\frac{-1}{\varepsilon(W,0)}\right]_{\text{MELF}} = \sum_n \frac{A_n}{E_n^2}\,\text{Im}\left[\frac{-1}{\varepsilon_{\text{D}}(W;E_n,\gamma_n)}\right] \quad (19.157)$$

where for the last equality we have used the relation

$$\text{Im}\left[\frac{-1}{\varepsilon_{\text{M}}(W,q=0;E_n,\gamma_n)}\right] = \text{Im}\left[\frac{-1}{\varepsilon_{\text{D}}(W;E_n,\gamma_n)}\right] = \frac{E_n^2\,\gamma_n W}{\left(E_n^2-W^2\right)^2 + (W\gamma_n)^2} \quad (19.158)$$

obtained from Equation (19.95). Note that Equation (19.158) is identical to the corresponding Ritchie-Howie expression at $q = 0$. Thus, the application of the MELF model also requires a Drude parameterization of the optical data.

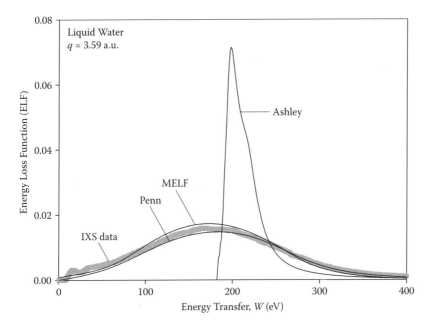

FIGURE 19.20

The experimental energy loss function of liquid water at the highest transferred momentum measured is compared with the Penn, Ashley, and MELF models.

However, unlike the Ritchie-Howie model, the advantage of the MELF model is that the extension of the optical data ELF to $q \neq 0$ is provided automatically through the analytic properties of the Mermin dielectric function. Thus, similar to the Penn model, no further approximations are needed for dispersion. However, the numerical computation of the ELF by Equation (19.156) is less cumbersome than by the integral form of Equation (19.148) of the Penn model. In Figure 19.20 it is shown that the MELF model can reproduce well the IXS data over the Bethe ridge to quality similar to that of the Penn model, whereas the Ashley model predicts a sharp peak that resembles that of the extended Drude model with the RPA dispersion (see Figure 19.19).

19.6.5 Hybrid Models

19.6.5.1 Liljequist Model

Liljequist (1983, 1985) proposed an extension algorithm that, in its initial form, does not require an analytic parameterization of the optical data. The starting point is the following expansion:

$$\frac{df(W, q = 0)}{dW} = \sum_n f_n \delta(W - W_n) \tag{19.159}$$

where f_n are the optical oscillator strengths associated with the excitation energy W_n and $\delta(W - W_n)$ are the so-called optical oscillators. Extension of Equation (19.159) to $q \neq 0$ to obtain the GOS may be accomplished in a straightforward manner according to

$$\frac{df(W,q)}{dW} = \sum_n f_n \delta(W - F_n(q)) \tag{19.160}$$

where $\delta(W - F_n(q))$ defines the so-called δ-oscillator with $F_n(q = 0) \equiv W_n$. To obtain a form suitable for use within an optical data model, W_n is replaced by the continuous excitation variable W' and Equation (19.160) is transformed to[*]

$$\frac{df(W,q)}{dW} = \int_0^\infty \left[\frac{df(W',q \approx 0)}{dW'} \right]_{\text{exp}} \delta(W - F(W',q))dW' \tag{19.161}$$

where $[df(W',q = 0)/dW']_{\text{exp}}$ denotes the experimental optical data for the OOS. Using now the proportionality relation between the GOS and the ELF, we can recast Equation (19.161) as

$$\text{Im}\left[\frac{-1}{\varepsilon(W,q)} \right] = \int_0^\infty \text{Im}\left[\frac{-1}{\varepsilon(W',q \approx 0)} \right]_{\text{exp}} \frac{W'}{W} \delta(W - F(W',q))dW' \tag{19.162}$$

with $F(W',q = 0) \equiv W'$. Thus, the extension of the optical data ELF to $q \neq 0$ in the Liljequist model is made via the delta-function $\delta(W - F(W',q))$. The determination of $F(W',q)$ follows an early categorization of Bohr whereby inelastic collisions are divided into two categories: (1) resonant-like interactions with bound electrons, corresponding to small momentum transfer, and (2) binary-like interactions with large momentum transfer, whereby the target electrons can be considered free (and at rest). Then, the $F(W',q)$ of Equation (19.162) takes the values

$$F(W',q) = \begin{cases} W' & \text{for} \quad Q \leq W' \\ Q & \text{for} \quad Q > W' \end{cases} \tag{19.163}$$

where we recall that $Q = q^2/2m$. Inserting Equation (19.163) into Equation (19.162) we obtain

$$\text{Im}\left[\frac{-1}{\varepsilon(W,q)} \right]_{\text{Liljequist}} = \text{Im}\left[\frac{-1}{\varepsilon(W,q \approx 0)} \right]_{\text{exp}} \Theta(W - Q)$$

$$+ \frac{\delta(W - Q)}{W} \int_0^Q W' \text{Im}\left[\frac{-1}{\varepsilon(W',q \approx 0)} \right]_{\text{exp}} dW' \tag{19.164}$$

---

[*] The present formulation follows Mayol and Salvat (1990), who have revisited Liljequist's model in the context of the optical data method.

Using the partial f-sum rule for the ELF, the integral on the right-hand side of Equation (19.164) becomes

$$\int_0^Q W' \text{Im} \left[\frac{-1}{\varepsilon(W', q \approx 0)} \right]_{\exp} dW' = \frac{\pi}{2} E_{pl}^2 \frac{Z_{eff}(Q)}{Z} \qquad (19.165)$$

with $Z_{eff}(Q)$ the effective number of target electrons that participate in collisions with zero momentum transfer ($q \approx 0$) and with energy transfer up to Q. Inserting Equation (19.165) into Equation (19.164) we obtain the more compact expression:

$$\text{Im} \left[\frac{-1}{\varepsilon(W, q)} \right]_{\text{Liljequist}} = \text{Im} \left[\frac{-1}{\varepsilon(W, q \approx 0)} \right]_{\exp} \Theta(W - Q) + \frac{\pi}{2} E_{pl}^2 \frac{Z_{eff}(Q)}{Z} \frac{\delta(W - Q)}{W}$$

$$(19.166)$$

The two terms in the right-hand side of Equation (19.166) represent the two types of collisions in the Bohr categorization. Specifically, the first term represents the soft or distant collisions with momentum transfer small enough so that $Q < W$, while the second term corresponds to the hard or close collisions with $W = Q$. In the former case the ELF is replaced by its optical limit, whereas in the latter case the ELF is replaced by a binary peak of zero width weighted by the portion of the zero momentum transfer f-sum rule for energy transfer up to Q. Apparently, for $Q > W$ the ELF of Equation (19.166) vanishes.

The Liljequist model is more suitable for core electron excitations than valence electron excitations in view of its very simplistic assumption of a constant ELF in the region $0 < Q < W$—i.e., of the approximation $\text{Im}[-1/\varepsilon(W, q)] = \text{Im}[-1/\varepsilon(W, q = 0)]$.

19.6.5.2 Two-Mode Model

Salvat and coworkers (Fernandez-Varea et al. 1992, 1993) have proposed the so-called two-mode model that is meant to reproduce the results of the Lindhard theory more closely than the one-mode approximation (used by Ashley) while also avoiding the computational burden of working with the full Lindhard dielectric function (as in the Penn model). The two-mode model starts from the following generalization of Equation (19.161) for the GOS:

$$\frac{df(W, q)}{dW} = \int_0^\infty \left[\frac{df(W', q = 0)}{dW'} \right]_{\exp} F(W, q; W') dW' \qquad (19.167)$$

where $F(W, q; W')$ is the extension algorithm with the property $F(W, q = 0; W') \equiv \delta(W - W')$. Taking advantage of the proportionality relation between the GOS and the ELF, we can recast Equation (19.167) in terms of the ELF as follows:

$$\text{Im}\left[\frac{-1}{\varepsilon(W,q)}\right] = \int_0^\infty \text{Im}\left[\frac{-1}{\varepsilon(W',q\approx 0)}\right]_{\text{exp}} \frac{W'}{W} F(W,q;W')\,dW' \quad (19.168)$$

It is straightforward to show that Equation (19.168) yields the Penn, Ashley, and Liljequist models following a suitable choice for $F(W,q;W')$:

$$F(W,q;W') = \begin{cases} F_{\text{Ashley}}(W,q;W') = \delta(W-(W'+Q)) \\ F_{\text{Liljequist}}(W,q;W') = \delta(W-W')\Theta(W'-Q) + \delta(W-Q)\Theta(Q-W') \\ F_{\text{Penn}}(W,q;W') = (2/\pi)(W/W'^2)\text{Im}[-1/\varepsilon_L(W,q;W')] \end{cases}$$

$$(19.169)$$

where $\varepsilon_L(W,q;W')$ is the Lindhard dielectric function with $W' = E_{pl}$.

To overcome the computational difficulties of adopting the full Lindhard dielectric function (as done in the Penn model), the two-mode approximation reads:

$$F_{\text{two-mode}}(W,q;W') \equiv [1-g(Q)]\delta(W-W'(Q)) + g(Q)\delta(W-Q) \quad (19.170)$$

where

$$W'(Q) = E_{pl} + BQ \quad (19.171)$$

and

$$g(Q) = \text{Min}\left\{1, AQ^3\left[E_{pl}^2(E_{pl}+Q)\right]^{-1}\right\} \quad (19.172)$$

with $A \equiv A(r_s)$ and $B \equiv B(r_s)$. The use of Equations (19.170) to (19.172) is meant to account in a simplified way for the most important aspect of the Lindhard dielectric function. Specifically, the first term in the right-hand side of Equation (19.170) corresponds to plasmon excitations, whereas the second term corresponds to single-electron excitations. For large Q, $g(Q) \approx 1$ and the plasmon contribution disappears, whereas that of the single-electron excitations leads to a sharply peaked Bethe ridge (i.e., of zero width). On the other hand, for small Q, $g(Q) \to 0$ roughly as Q^3 and the plasmon mode becomes dominant, while the single-electron contribution gradually disappears. The forms of $W'(Q)$, $g(Q)$, $A(r_s)$, and $B(r_s)$ are not given by the model but have to be determined by independent calculations. The detailed expressions can be found in the original papers (Fernandez-Varea et al. 1992, 1993). Importantly, the extension algorithm given by Equations (19.170) to (19.172) allows the inelastic scattering cross section to be calculated analytically. On the other hand, the two-mode model shares the same problems with the Ashley (and Penn) models with respect to threshold effects; thus, it is generally unsuited for core electron excitations from inner shells with large binding energies. To that end, the suggestion is made (Fernandez-Varea et al. 1993) to combine the two-mode model with the Liljequist model by considering two different excitation regions with different extension algorithms as follows:

$$F(W,q;W') = \begin{cases} F_{\text{two-mode}}(W,q;W') & \text{for } W' < W_s \\ F_{\text{Liljequist}}(W,q;W') & \text{for } W' > W_s \end{cases} \tag{19.173}$$

where W_s is the energy that makes the extension algorithm to switch from the low-Q regime of plasmon-like excitations to the high-Q of single-electron excitations. In practice the value of W_s is set equal to the smallest core shell binding energy.

References

Abril I, Denton CD, de Vera P, Kyriakou I, Emfietzoglou D, Garcia-Molina R. 2010. Effect of the Bethe surface description on the electronic excitations induced by energetic proton beams in liquid water and DNA. *Nucl. Instrum. Meth. B* 268:1763–1767.

Abril I, Garcia-Molina R, Denton CD, Perez-Perez JF, Arista N. 1998. Dielectric description of wakes and stopping powers in solids. *Phys. Rev. A* 58 357–366.

Ashley JC. 1988. Interaction of low-energy electrons with condensed matter: stopping powers and inelastic mean free paths from optical data. *J. Electron Spectrosc. Relat. Phenom.* 46: 199–214.

Ashley JC. 1990. Energy loss rate and inelastic mean free path of low-energy electrons and positrons in condensed matter. *J. Electron Spectrosc. Relat. Phenom.* 50: 323–334.

Bohm D, Pines D. 1953. A collective description of electron interactions. III. Coulomb interactions in a degenerate electron gas. *Phys. Rev.* 92: 609–625.

Chan WF, Cooper G, Brion CE. 1993. The electronic spectrum of water in the discrete and continuum regions. Absolute optical oscillator strengths for photoabsorption (6–200 eV). *Chem. Phys.* 178: 387–400.

Ding Z-J, Shimizu R. 1989. Inelastic collisions of kV electrons in solids. *Surf. Sci.* 222: 313–331.

Dingfelder M, Hantke D, Inokuti M, Paretzke HG. 1998. Electron inelastic-scattering cross sections in liquid water. *Radiat. Phys. Chem.* 53: 1–18.

Dingfelder M, Inokuti M. 1999. The Bethe surface of liquid water. *Radiat. Environ. Biophys.* 38: 93–96.

Dingfelder M, Inokuti M, Paretzke HG. 2000. Inelastic-collision cross sections of liquid water for interactions of energetic protons. *Radiat. Phys. Chem.* 59: 255–275.

Egerton RF. 1996. *Electron Energy-Loss Spectroscopy in the Electron Microscope*, 2nd ed. New York: Plenum Press.

Emfietzoglou D, Abril I, Garcia-Molina R, Petsalakis ID, Nikjoo H, Kyriakou I, Pathak A. 2008. Semi-empirical dielectric descriptions of the Bethe surface of the valence bands of condensed water. *Nucl. Instrum. Meth. B* 266: 1154–1161.

Emfietzoglou D, Cucinotta F, Nikjoo H. 2005. A complete dielectric response model for liquid water: A solution of the Bethe ridge problem *Radiat. Res.* 164: 202–211.

Emfietzoglou D, Garcia-Molina R, Kyriakou I, Abril I, Nikjoo H. 2009. A dielectric response study of the electronic stopping power of liquid water for energetic protons and a new I-value for water. *Phys. Med. Biol.* 54: 3451–3472.

Emfietzoglou D, Nikjoo H. 2005. The effect of model approximations on single-collision distributions of low-energy electrons in liquid water. *Radiat. Res.* 163: 98–111.

Fano U. 1956. Atomic theory of electromagnetic interactions in dense materials. *Phys. Rev.* 103: 1202–1218.

Fano U. 1963. Penetration of protons, alpha particles, and mesons. *Annu. Rev. Nucl. Sci.* 13: 1–66.

Feibelman PJ. 1973. Spatial variation of the electron mean free path near a surface. *Surf. Sci.* 36: 558–568.

Fermi E. 1940. The ionization loss of energy in gases and in condensed materials. *Phys. Rev.* 57: 485–493.

Fernandez-Varea JM, Mayol R, Liljequist D, Salvat F. 1993. Inelastic scattering of electrons in solids from a generalized oscillator strength model using optical and photoelectric data. *J. Phys. Condens. Matter* 5: 3593–3610.

Fernandez-Varea JM, Mayol R, Salvat F, Liljequist D. 1992. A comparison of inelastic scattering models based on a δ-function representation of the Bethe surface. *J. Phys. Condens. Matter* 4: 2879–2890.

Fernandez-Varea JM, Salvat F, Dingfelder M, Liljequist D. 2005. A relativistic optical-data model for inelastic scattering of electrons and positrons in condensed matter. *Nucl. Instrum. Methods B* 229: 187–218.

Giuliani GF, Vignale G. 2005. *Quantum Theory of the Electron Liquid*. Cambridge: Cambridge University Press.

Hamm RN, Wright HA, Ritchie RH, Turner JE, Turner TP. 1975. Monte Carlo transport of electrons through liquid water. In *5th Symposium on Microdosimetry*. Brussels: EUR-5452 EC, pp. 1037–1050.

Hayashi H, Watanabe N, Udagawa Y, Kao CC. 2000. The complete optical spectrum of liquid water measured by inelastic x-ray scattering. *Proc. Natl. Acad. Sci. USA* 97: 6264–6266.

Hedin L, Lundqvist S. 1969. Effects of electron-electron and electron-phonon interactions on the one-electron states of solids. *Sol. Stat. Phys.* 23: 1–181.

Heller JM, Hamm RN, Birkhoff RD, Painter LR. 1974. Collective oscillation in liquid water. *J. Chem. Phys.* 60: 3483–3486.

Howie A, Stern RM. 1972. The optical potential in electron diffraction. *Z. Naturf. A* 27: 382–389.

Hubbard J. 1955. The dielectric theory of electronic interactions in solids. *Proc. Phys. Soc. Lond. A* 68: 976–986.

Hubbard J. 1958. The description of collective motion in terms of many-body perturbation theory. II. The correlation energy of a free-electron gas. *Proc. Roy. Soc. Lond. A* 243: 336–352.

Inagaki T, Hamm RN, Arakawa ET. 1974. Optical and dielectric properties of DNA in the extreme ultraviolet. *J. Chem. Phys.* 61: 4246–4250.

Inokuti M. 1971. Inelastic collisions of fast charged particles with atoms and molecules: The Bethe theory revisited. *Rev. Mod. Phys.* 43: 297–347.

Kaplan IG, Miterev AM. 1987. Interaction of charged particles with molecular medium and track effects in radiation chemistry. *Adv. Chem. Phys.* LXVIII: 255–386.

Kleinman L. 1967. New approximation for screened exchange and the dielectric constant of metals. *Phys. Rev.* 160: 585–590.

Kliewer KL, Fuchs R. 1969. Lindhard dielectric function with a finite electron lifetime. *Phys. Rev.* 181: 552–558.

Kobayashi K. 1983. Optical spectra and electronic structure of ice. *J. Phys. Chem.* 87: 4317–4321.

Kuhr J-C, Fitting H-J. 1999. Monte Carlo simulation of electron emission from solids. *J. Electron Spectr. Related Phenom.* 105: 257–273.

Langreth DC. 1969. Approximate screening functions in metals. *Phys. Rev.* 181: 753–762.

Liljequist D. 1983. A simple calculation of inelastic mean free path and stopping power for 50 eV–50 keV electrons in solids. *J. Phys. D Appl. Phys.* 16: 1567–1582.

Liljequist D. 1985. Simple generalized oscillator strength density model applied to the simulation of keV electron-energy-loss distributions. *J. Appl. Phys.* 57: 657–665.

Liljequist D. 2008. A study of errors in trajectory simulation with relevance for 0.2–50 eV electrons in liquid water. *Radiat. Phys. Chem.* 77: 835–853.

Lindhard J. 1954. On the properties of a gas of charged particles. *Mat. Fys. Medd. Dan. Vid. Selsk* 28(8): 1–57.

Mayol R, Salvat F. 1990. Cross sections for K-shell ionization by electron impact. *J. Phys. B At. Mol. Opt. Phys.* 23: 2117–2130.

Mermin ND. 1970. Lindhard dielectric function in the relaxation-time approximation. *Phys. Rev. B* 1: 2362–2363.

Nozieres P, Pines D. 1959. Electron interaction in solids. Characteristic energy loss spectrum. *Phys. Rev.* 113: 1254–1267.

Pauly N, Tougaard S. 2009. Determination of the effective surface region thickness and of Begrenzungs effect. *Surf. Sci.* 603: 2158–2162.

Penn DR. 1987. Electron mean-free-path calculations using a model dielectric function. *Phys. Rev. B* 35: 482–486.

Pines D. 1963. *Elementary Excitations in Solids*. New York: Benjamin.

Powell CJ. 1968. Inelastic scattering of kilovolt electrons by solids and liquids: Determination of energy losses, cross sections, and correlations with optical data. *Health Physics* 13: 1265–1275.

Powell CJ. 1974. Attenuation lengths of low-energy electrons in solids. *Surf. Sci.* 44: 29–46.

Powell CJ, Jablonski A. 1999. Evaluation of calculated and measured electron inelastic mean free paths near solid surfaces. *J. Phys. Chem. Ref. Data* 28: 19–62.

Ritchie RH. 1957. Plasma losses by fast electrons in thin films. *Phys. Rev.* 106: 874–881.

Ritchie RH. 1959. Interaction of charged particles with a degenerate Fermi-Dirac electron gas. *Phys. Rev.* 114: 644–654.

Ritchie RH, Hamm RN, Turner JE, Wright HA. 1978. The interaction of swift electrons with liquid water. In *6th Symposium on Microdosimetry*. Luxemburg: Harwood Academic, pp. 345–354.

Ritchie RH, Hamm RN, Turner JE, Wright HA, Bolch WE. 1991. Radiation interactions and energy transport in the condensed phase. In *Physical and Chemical Mechanisms in Molecular Radiation Biology*. New York: Plenum Press, pp. 99–136.

Ritchie RH, Howie A. 1977. Electron excitation and the optical potential in electron microscopy. *Philos. Mag.* 36: 463–481.

Ritsko JJ, Brillson LJ, Bigelow RW, Fabish TJ. 1978. Electron energy loss spectroscopy and the optical properties of polymethylmethacrylate from 1 to 300 eV. *J. Chem. Phys.* 69: 3931–3939.

Rossi HH, Zaider M. 1996. *Microdosimetry and Its Applications.* Springer-Verlag, Berlin Germany.

Schattschneider P. 1986. *Fundamentals of Inelastic Electron Scattering.* Springer-Verlag, Wein Austria.

Segui S, Dingfelder M, Fernandez-Varea JM, Salvat F. 2002. The structure of the Bethe ridge. Relativistic Born and impulse approximation. *J. Phys. B At. Mol. Opt. Phys.* 35: 33–53.

Tanuma S, Powell CJ, Penn DR. 1993a. Use of sum rules on the energy-loss function for the evaluation of experimental optical data. *J. Electron Spectrosc. Relat. Phenom.* 62: 95–109.

Tanuma S, Powell CJ, Penn DR. 1993b. Calculations of electron inelastic mean free paths. V. Data for 14 organic compounds over the 50–2000 eV range *Surf. Interf. Anal.* 21: 165–176.

Tanuma S, Powell CJ, Penn DR. 2003. Calculation of electron inelastic mean free paths (IMFPs). VII. Reliability of the TPP-2M IMFP predictive equation. *Surf. Interf. Anal.* 35: 268–275.

Watanabe N, Hayashi H, Udagawa Y. 1997. Bethe surface of liquid water determined by inelastic x-ray scattering spectroscopy and electron correlation effects. *Bull. Chem. Soc. Jpn.* 70: 719–726.

Watanabe N, Hayashi H, Udagawa Y. 2000. Inelastic x-ray scattering study on molecular liquids. *J. Phys. Chem. Solids* 61: 407–409.

Yubero F, Sanz JM, Ranskov B, Tougaard S. 1996. Model for quantitative analysis for reflection-electron-energy-loss spectra: Angular dependence. *Phys. Rev. B* 53: 9719–9727.

Zaider M. 1991. Charged particle transport in the condensed phase. In *Physical and Chemical Mechanisms in Molecular Radiation Biology.* New York: Plenum Press, pp. 137–162.

Section IV

20

Questions and Problems

1. Define the following terms and complete the equations assuming XY, X and Y are molecules and atoms.

   a. Direct ionization   XY $\rightarrow$

   b. Excitation   XY $\rightarrow$

   c. Superexcitation   XY $\rightarrow$ XY'

   d. Autoionization   XY' $\rightarrow$

   e. Autoionization of water $2H_2O$ $\rightarrow H_3O^+ + OH^-$

   f. Dissociation   XY' $\rightarrow$

                             XY^* $\rightarrow$

                             $H_3O^+ + e^-$ $\rightarrow$

                             $CH_3^+ + e^-$ $\rightarrow$

                             $H_2O^+ + e^-$ $\rightarrow$

   g. Internal conversion (IC)   XY' $\rightarrow$

   h. Electron capture (EC) $e^- + p^+$ $\rightarrow$

   i. How are Auger electrons produced?

   j. Fluorescence XY^* $\rightarrow XY + h\nu$

   k. Radical recombination   2X $\rightarrow X_2$

   l. Addition   X + XY $\rightarrow$

   m. Excimer formation $XY^* + XY$ $\rightarrow$

   n. Electron attachment $e^- + S$ $\rightarrow S^-$

   o. Reaction   XY + S $\rightarrow$

2. Describe the radiative and nonradiative transitions for an excited He atom.

3. Describe how Auger electrons are quantified.

4. What are the sequences of events following interaction of electromagnetic radiation with matter, and the type of interactions that occur?

5. Describe the decay scheme of 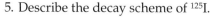 ^{125}I.

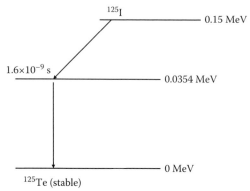

6. Sketch diagrams for the following interactions:
   a. Photoelectric
   b. Compton
   c. Rayleigh
   d. Pair production
   e. Elastic and inelastic
   f. Bremsstrahlung
   g. Inelastic
   h. Annihilation

7. The 35 W LPS lamp produces a virtually monochromatic light averaging a wavelength of 589.3 nm. A person is standing 10 m away from this lamp with a camera taking pictures of the lamp. The aperture of the camera is set at 4, the focal length is 100 mm, and the shutter speed is 1/100 s. How many photons enter the camera during the exposure?

8. At a wavelength of light greater than 650.0 nm the emission of photoelectrons from a surface ceases. The surface is irradiated with a wavelength of 400 nm. What is the maximum energy of the electrons emitted from the surface?

9. Show that the total cross section for Thomson scattering is $\frac{8}{3}\pi r_e^2$.

10. A low-energy photon is Thomson scattered at an angle of 30° with respect to its incident direction.
    a. Calculate the differential cross section $\frac{d\sigma}{d\Omega}$ [b sr^{-1}] of an electron.
    b. Calculate the differential cross section $\frac{d\sigma}{d\theta}$ [b rad^{-1}] of an electron.

11. A 30 keV photon incident on water molecule is coherent scattered.

Use the following table of the atomic form factors $F(Z, v)$ for a logarithmic interpolation.

v	$F(1,v)$	$F(8,v)$
0.2	0.48078	5.6197
0.25	0.34974	4.8047
1.5	0.001494	0.9961
2.0	0.000489	0.6720

a. Calculate $\frac{d\sigma}{d\theta}$ [b rad⁻¹] at 10°.

b. Calculate $\frac{d\sigma}{d\theta}$ [brad⁻¹] at 90°.

12. The binding energies of Pb shells are listed in the table.

K	L_I	L_{II}	L_{III}
88.001	15.870	15.207	13.044

A 100 keV photon is absorbed by Pb atom. What is the energy of the photoelectron ejected from each shell?

13. Show that the conservation of energy and momentum prevent the absorption of a photon by a free electron.

14. K-shell photoabsorption occurred at the surface of iron block irradiated by 100 keV photons. Photoelectrons with the energy of 92.9 keV were ejected. What is the K absorption edge of iron?

15. The effective atomic numbers for the muscle and bone are 7.64 and 12.31, respectively. Estimate the ratio of photoelectric absorption cross section for ~30 keV photons.

16. a. Photoabsorption cross sections for 30 keV photons are 38.2 b for Al atom and 4.32 b for O atom, respectively. Calculate the Z dependence of each cross section.

   b. Photoabsorption cross sections for 30 and 60 keV photons are 38.2 and 4.32 b for Al atom, respectively. Calculate the energy dependence of each cross section.

17. Photoelectrons are emitted with the energy of 100 eV when photons are incident on the hydrogen atom ($B = 13.6$ eV). Calculate the wavelength of incident photons.

18. What is the stopping potential to stop electron emission when a light of frequency 1.5×10^{15} s⁻¹ is incident on a tungsten surface (work function 4.52 eV)?

19. A K-L_2-L_2 Auger electron with the energy of 10 keV was emitted from an element with the K-shell binding energy of 15 keV. What is the binding energy of the L_2-shell and the energy of the $K_{\alpha 2}$ characteristic x-ray?

20. Show that the threshold photon wavelength for producing an electron-positron pair is 0.0012 nm.

21. An electron with 6×10^7 ms^{-1} and a positron with 1.5×10^8 ms^{-1} are emitted. What is the incident photon energy?

22. The energy of an electron was three times that of a positron generated by 10 MeV photons. What is the energy of the electron?

23. Derive Equation (7.4).

24. Derive Equation (7.5).

25. Derive Equation (7.7).

26. From Equation (7.7), show that the recoil angle of an electron is limited to the forward angles.

27. a. Show that the difference of the wavelength between the incident photon and the scattered photon is given by
$$\Delta\lambda = \lambda' - \lambda = \frac{h}{mc}(1 - \cos\theta).$$

   b. On the equation in (a), calculate the numerical value of h/mc, which is called the Compton wavelength.

28. A 1 MeV photon is Compton scattered at an angle of 60°. Calculate:
   a. The energy of the scattered photon
   b. The change in wavelength
   c. The angle of recoil of the electron
   d. The recoil energy of the electron

29. A 511 keV photon induced Compton scattering with an electron. What is the energy of the scattered photon when the electron is emitted at an angle of 45° with respect to the incident direction?

30. A photon is observed to be scattered at an angle 120° while the electron recoils at an angle 20° with respect to the incident direction.
   a. What is the incident photon energy?
   b. What is the frequency of the scattered photon?
   c. How much energy does the electron receive?

31. Compton scattering occurred for a 30 keV photon in water. Calculate the $\frac{d\sigma}{d\theta}$ at an angle 90°.

32. Incoherent scattering occurred for a 30 keV photon in water. Calculate the $\frac{d\sigma_{incoh}}{d\theta}$ at an angle 90°. Use the following table of the scattering functions $S(Z,v)$ for a logarithmic interpolation.

v	$S(1,v)$	$S(8,v)$
1.5	1.0	7.462
2.0	1.0	7.764

33. The linear attenuation coefficient of Fe for 1 MeV photon is 0.47 cm^{-1}. Calculate the atomic cross section.

34. The total cross section for 500 keV photons in Al is 3.78 b. Calculate the mass attenuation coefficient.

35. The atomic cross sections for 1 MeV photon interactions with hydrogen and oxygen are, respectively, 0.21 b and 1.69 b.

    a. Calculate the linear attenuation coefficient for water.

    b. Calculate the mass attenuation coefficient.

36. A narrow beam of 150 keV photons is directed normally at a slab of copper. The fraction of the photons penetrating the slab was 0.6. What is the thickness of the slab? (Attenuation coefficient = 0.222 cm^2 g^{-1} and density = 8.93 g cm^{-3}.)

37. A narrow beam of 200 keV photons is directed normally at a 7 mm copper slab. The fraction of the photons penetrating the slab was 0.38. What is the half-value layer of copper?

38. A narrow beam of 150 keV photons is normally incident on a copper absorber of thickness 3 mm. The total number of photons is 1×10^6. The mass attenuation coefficient is 0.2217 cm^2 g^{-1} and the density of copper is 8.93 g cm^{-3}. What is the energy dissipated in the absorber?

39. Derive Equation (7.22).

40. Calculate the mean excitation energy I for water.

41. Calculate $F^-(\tau)$ given by Equation (8.5) for a 500 keV electron.

42. Calculate $F^+(\tau)$ given by Equation (8.6) for a 500 keV positron.

43. Calculate the collision stopping power of Al ($I = 163$ eV) for 500 keV electrons assuming $\delta = 0$.

44. Calculate the collision stopping power of Al for 500 keV positrons assuming $\delta = 0$.

45. What is the ratio of the collisional and radiative stopping powers of Fe for electrons of energy:

    a. 100 keV

    b. 10 MeV

    c. 100 MeV

46. Using Equation (8.13), estimate the radiation yield for 20 MeV electrons in:

    a. Water

    b. Fe

    c. Pb

47. Using Equation (8.15), estimate the range in water for electrons of energy:

    a.   100 keV

    b.   4 MeV

48. Derive Equation (8.17).

49. Calculate G in Equation (10.5) for a 1 MeV electron for $\Delta = 100$ eV.

50. Using the answer of Question 49, calculate restricted collision stopping power of water ($I = 75$ eV) assuming $\delta = 0$.

51. Derive Equation (11.26).

52. Obtain A, B, and C in Equation (11.32) for a cylinder with the radius r centered at $(0, 0, d)$ and with the infinite length along the y-axis.

53. Estimate the total cross section for ionization at $T = 100$ eV using Equation (13.10).

54. Calculate the collision stopping power of a 1 keV electron for water using Equation (13.21). Read Figure 13.7 to obtain the cross sections for ionization and excitation. Use the mean energy transfers, $\bar{\varepsilon} = 29$ eV, $\bar{B} = 17$ eV, $\bar{E}_a = 13.5$ eV.

55. Derive Equation (10.1).

56. Derive Equation (8.2).

57. Show the maximum energy of electrons ejected by proton impact, given by $4T/\lambda$, in which T is the kinetic energy of proton and λ is the mass ratio m_p/m_e.

58. Calculate the maximum energy that a 1 MeV proton can transfer to an electron in a single collision.

59. Calculate the collision stopping power of water ($I = 75$ eV) for 8 MeV α-particles using Equation (9.1).

60. Calculate $d\sigma_{el}/d\Omega$ (b sr^{-1}), a target molecule at 5° for 5 MeV α-particles in water using Equation (9.13).

61. Show that the common LET unit, keV $\mu\mathrm{m}^{-1}$, is equal to 10 MeV cm^{-1}.

62. Confirm the relationship Equation (9.19) from Figure 9.4.

63. Using Equations (14.17) and (14.18), calculate the electronic stopping power at 100 keV protons for water.

    a.   Use Figure 14.3 to obtain the cross sections for ionization and excitation.

    b.   Use Figure 14.6 to obtain the charge transfer cross sections.

    c.   Using the answer for (b), estimate the fractions of proton and neutral hydrogen in the equilibrium charge state.

    d. Use Figure 14.7 to obtain the mean energy transfers for ionization, excitation, and charge transfer processes.

    e. Using these answers, estimate S_e for water.

64. What is the kinetic energy of electrons with the same speed of 100 MeV protons?

65. Use Figure 13.7 to obtain the ionization cross section for 18.4 MeV protons and compare the data presented in Figure 16.3.

66. State the two main assumptions of the plane wave Born approximation (PWBA).

67. Calculate the validity limits of the PWBA for incident electrons and protons in water medium based on Equation (19.10) assuming an effective atomic number for water of 7.5. How do the values compare with the practically adopted limits?

68. The factorization of Equation (19.11) into a particle-dependent factor and a material-dependent factor (the dynamic structure factor) has an important practical implication. What is it?

69. What is the practical significance of the Bethe asymptotic expansion of Equation (19.50)?

70. The Bethe stopping power formula, Equation (19.56), is based on two approximations. Which are they?

71. What is the origin of the inner shell effects in the Bethe theory?

72. What is the zero-energy density effect in Bethe's stopping power theory?

73. What is a free-electron-like material?

74. How does a real material differ from the HEG in terms of its plasmon characteristics?

75. How does the dielectric function (and the ELF) reflect the single-electron and plasmon excitations of a system?

76. What is the relation between $\mathrm{Im}(-1/\varepsilon)$ and $\mathrm{Im}(\varepsilon)$ under screening, antiscreening, and no screening conditions?

77. State the main assumption of the homogeneous electron gas (HEG).

78. What is the underlying assumption of the random phase approximation (RPA) to the dielectric response of HEG?

79. What is the Landau damping mechanism and why is there a critical momentum-transfer cutoff associated with it?

80. What happens to the Lindhard energy loss function at values of momentum transfer above the Landau cutoff?

81. What is the difference between the Lindhard and Mermin dielectric functions?

82. How do the Lindhard and Mermin dielectric functions relate to the Drude model?

83. The many-body local field correction (LFC) is meant to account for two effects that are neglected in the RPA. What are they?

84. What is the effect of the LFC upon the plasmon dispersion relations, and how do these relations change upon the use of a static LFC (SLFC)?

85. Do you expect an optical data model to be more reliable at high- or low-incident particle energies and why?

86. Which of the following models predict a broadened energy loss function over the Bethe ridge? (a) Extended-Drude model with RPA dispersion, (b) Liljequist model, (c) Penn model, (d) Ashley model, or (e) MELF model.

87. Under which approximation does the Penn model reduce to the Ashley model?

88. What is the main problem of the Ritchie-Howie model that the MELF model overcomes?

89. Write a computer program to obtain 1,000 random numbers between 0 and 1. Draw a frequency distribution of these numbers.

90. Write a computer program to obtain 100 random chords in a circle, diameter d. Plot the chords.

91. Write a computer program to obtain random chords cutting a sphere, radius r, using the method of μ-randomness.

92. Write a computer program to calculate MFP for protons 1–1,000 keV.

93. Write a computer program to calculate oscillator strength for water in the range 10–100 eV photons.

   a. For <30 eV

   b. For 30 eV < E < 10 keV

Index